Engineering Edu
and Practice in
United States

Foundations of Our Techno-Economic Future

Committee on the Education and Utilization of the Engineer
Commission on Engineering and Technical Systems
National Research Council

NATIONAL ACADEMY PRESS
Washington, D.C. 1985

NATIONAL ACADEMY PRESS 2101 Constitution Avenue, NW Washington, DC 20418

NOTICE: The project that is the subject of this report was approved by the Governing Board of the National Research Council, whose members are drawn from the councils of the National Academy of Sciences, the National Academy of Engineering, and the Institute of Medicine. The members of the committee responsible for the project were chosen for their special competences and with regard to appropriate balance.

This report has been reviewed by a group other than the authors according to procedures approved by a Report Review Committee consisting of members of the National Academy of Sciences, the National Academy of Engineering, and the Institute of Medicine.

The National Research Council was established by the National Academy of Sciences in 1916 to associate the broad community of science and technology with the Academy's purposes of furthering knowledge and advising the federal government. The Council operates in accordance with general policies determined by the Academy under the authority of its congressional charter of 1863, which establishes the Academy as a private, nonprofit, self-governing membership corporation. The Council has become the principal operating agency of both the National Academy of Sciences and the National Academy of Engineering in the conduct of their services to the government, the public, and the scientific and engineering communities. It is administered jointly by both Academies and the Institute of Medicine. The National Academy of Engineering and the Institute of Medicine were established in 1964 and 1970, respectively, under the charter of the National Academy of Sciences.

First Printing, April 1985
Second Printing, July 1986

Library of Congress Catalog Card Number 85-60423
International Standard Book Number 0-309-03539-2

Printed in the United States of America

PREFACE AND ACKNOWLEDGMENTS

In the early 1980s the engineering profession was in ferment over its future. Soaring undergraduate enrollments coupled with faculty shortages threatened the quality of engineering education. At the same time, industry struggled to recruit adequate numbers of engineering graduates to meet the nation's growing needs. Clearly, these problems go beyond the university and the board room—in a society increasingly dependent on high technology they command national attention.

Despite engineering's crucial role in modern economic life, public debate on technology development and its impact on the national and global economies have not often included examination of the engineering profession per se. Prompted by concern over the health of the U.S. engineering endeavor, the National Science Foundation asked the National Research Council in 1980 to conduct a study of the state and the future of engineering education and practice in the United States.

The Committee on the Education and Utilization of the Engineer consisted of 26 members and 9 panels with more than 50 additional people drawn from business, industry, and education. These groups, which included all facets of engineering as well as other disciplines such as the social sciences and economics, met at regular intervals for two years to develop the findings and recommendations contained in this volume. One member of the committee was also the director of a two-year study of faculty shortages begun in 1981 by the American

Society for Engineering Education. (All study participants are listed in Appendix A.)

In this report and in several forthcoming companion reports (see Appendix B), the committee attempts to present a comprehensive view of how—and how well—the engineering community functions. This view is directed toward a wide and diverse audience: national leadership in both the public and private sectors, the nonengineering public, and of course, the broad engineering community itself.

Although the findings and recommendations of this report are meant to guide and inform this audience, it should be remembered that they are generic and thus cannot cover every situation. For example, some segments of society conclude that missed schedules, cost overruns, and technical shortcomings in engineering projects indicate a deficiency in engineering capability in this country. Yet because not all projects suffer from these difficulties, we surmise that the problem lies more in management effectiveness than in engineering capability. Thus we make no recommendations on what we perceive to be an individualized, organizational problem.

By the same token, each committee member must admit to forming conclusions based on insights from evidence that, if put to the test, would not have produced a ringing consensus. Hardly anyone involved in the give-and-take of the committee effort could escape learning new things and forming new judgments that, in turn, have become an important component of this report. For example, we are aware of intense pressures to modify the undergraduate engineering curriculum to include more subjects in the humanities, liberal arts, and social sciences as well as more technical and business courses, all within the confines of a sacrosanct four-year program. Arguments on all sides are unimpeachable but they are also mutually exclusive, and moving in favor of any one of them causes the root curriculum to suffer. The arguments could be reconciled in a plan for a preengineering undergraduate program followed by a professional school program, with the combination requiring more time to earn the first professional degree. However, because of objections to the extra costs of this approach and the expected reluctance on the part of students to extend their college program, the committee could not reach a consensus on this vexing problem.

The architects of this study predicted that it would be difficult if not impossible to complete a task of such scope in two years; the committee can now confirm this prediction. We hope to see our work become the first step in a continuing effort that will yield judgments and recommendations for which we could lay only the groundwork.

Support for this work has been provided by the National Science Foundation, the Department of the Air Force, the Department of the Army, the Department of Energy, the Department of the Navy, and the National Aeronautics and Space Administration. Additionally, assistance has been provided through grants from the Eastman Kodak Company, Exxon Corporation, the General Electric Company, the IBM Corporation, the Lockheed Corporation, the Monsanto Company, and the Sloan Foundation. We thank all of these groups for their support and encouragement.

The committee expresses its appreciation to all the participants in the study —panel members, consultants, and staff—for their dedicated efforts in carrying out the extensive undertakings required in its conduct. The efforts of D. D. Wyatt in the early stages should not be overlooked. In particular, the committee and staff thank Courtland S. Lewis for his valuable contributions to the preparation of this report. As chairman, I appreciate greatly the wisdom and cooperation of all contributors, but most especially William H. Michael, Jr., and David C. Hazen.

JERRIER A. HADDAD

CHAIRMAN

Committee Members

JERRIER A. HADDAD, Chairman, (IBM, Ret.)
GEORGE S. ANSELL, Dean of Engineering, Rensselaer Polytechnic Institute (now President, Colorado School of Mines)
JORDAN J. BARUCH, President, Jordan J. Baruch Associates
ERICH BLOCH, Vice-President, IBM Corporation (now Director, National Science Foundation)
DENNIS CHAMOT, Associate Director, Department for Professional Employees, AFL/CIO
EDMUND T. CRANCH, President, Worcester Polytechnic Institute
DANIEL C. DRUCKER, Dean of Engineering, University of Illinois at Urbana (now Graduate Research Professor of Engineering Sciences, University of Florida at Gainesville)
FRED W. GARRY, Vice-President, Corporate Engineering and Manufacturing, General Electric Company
JOHN W. GEILS, Director of AAES/ASEE Faculty Shortage Project (AT&T, Ret.)
AARON J. GELLMAN, President, Gellman Research Associates, Inc.
HELEN GOULDNER, Dean, College of Arts and Science, Professor of Sociology, University of Delaware
JOHN D. KEMPER, Professor, Department of Mechanical Engineering, University of California at Davis
EDWARD T. KIRKPATRICK, President, Wentworth Institute of Technology

CONTENTS

Engineering Education and Practice in the United States

Foundations of Our Techno-Economic Future

EXECUTIVE SUMMARY

BROAD ISSUES IN ENGINEERING

The Committee on the Education and Utilization of the Engineer has conducted a broad study aimed at achieving a comprehensive understanding of engineering in the United States and assessing its capacity to meet present and future challenges. Over a two-year period the committee addressed a great many specific questions relating to the characteristics and functioning of engineers. As a result, its findings in these areas are numerous and detailed. Apart from these detailed findings, the committee also addressed broad questions that cut across the various areas of study, and for that reason they do not directly reflect the organizational plan of the report itself. By addressing the following five broad issues, this summary attempts to convey the essence of the fun report and its findings.

IS THE ENGINEERING EDUCATIONAL SYSTEM HEALTHY?

When the committee began its work in 1982, there was a widespread perception of crisis in engineering education. Accordingly, the committee examined this situation very closely. Its findings indicate that the situation was indeed critical in many schools, primarily because of a tremendous increase in enrollments in the face of faculty shortages. In many schools the capacity to cope was and still is being strained severely, but the educational system is managing (albeit with varying degrees of strain from school to school.). Simply getting by is not satisfactory, however, and it is not acceptable.

The committee believes that it is not productive to debate whether the problems in engineering education are of crisis proportion. But there are problems in engineering colleges that vary in intensity depending on the individual situation. The faculty shortage is proving particularly hard to redress because too few students choose to go into graduate study for the Ph.D and because too few of these have wanted to take faculty positions. Increases in current doctoral enrollments provide hope for at least some improvement in this area—especially because undergraduate enrollments seem to have leveled off and because schools are now making stronger efforts to improve faculty compensation and the academic work environment. Nevertheless, the problem is still far from solved in many institutions.

Among other concerns, over 40 percent of the anticipated new Ph.D. graduates will be foreign students on temporary visas and thus probably will not be available to help meet the faculty shortage. In some schools, laboratory equipment is obsolete and physical plants and facilities have deteriorated—problems that grow more severe with each passing school term and with each advance in science and technology. There is also the continuing difficulty of providing a broad education in engineering fundamentals, a degree of specialized knowledge in a certain field, a general education, and communication and technical managerial skills in four years.

However, the committee notes that the public's regard for engineering education has risen in recent years (as seen, for example, in increased appropriations by various state legislatures) and recognizes that the quality of engineering students and graduates alike has been very high. In addition, educational technology and continuing education offer increasingly powerful and affordable means to alleviate some of the existing problems.

These views are not universally shared. Some respected members of the engineering educational community feel that the problems remain dangerously severe and that improvements are merely cosmetic. They are concerned that the overall level of technical education in this country will not sustain the nation's leadership in the face of worldwide growth in technical competency.

The committee recognizes that the future is uncertain and that international competition will increasingly test the strength of engineering education. Although the engineering educational system does show some signs of recovering from the severe problems it has experienced, additional efforts and support on the part of schools, industry, and government are required in many areas to improve the health of the system.

HOW COMPETENT ARE ENGINEERS IN THE UNITED STATES?

In light of a number of highly publicized engineering failures in recent years, it is pertinent to ask whether the quality of U.S. engineering is good enough to protect public health and safety and to achieve national goals.

The committee found a widespread opinion within industry that the competence of recently graduated engineers is higher than ever before. There is no evidence of a serious flaw in the basic technical education of entry-level engineers. On the contrary, the new engineers have strong analytical skills and an excellent theoretical base in engineering science. However, most companies find that the contemporary graduate lacks the ability to step into a job and become immediately productive. Although this is not a new problem, it has been exacerbated by the trend toward fewer design or practice courses. Often, additional training of six months to a year or more is required to acclimate the new engineer properly to the requirements of the job. Some aspects of this in-house training are simply specific to a given company (procedures, special products and terminologies, etc.) and as such are unavoidable. Other aspects are industry-specific, or involve bringing the engineer up to the state of the art in the industry. Offering this training is a particular problem for smaller companies because of its cost.

Another element of the problem is that to make the transition from a high school graduate to a competent practicing engineer requires more than just the acquisition of technical skills and knowledge. It also requires a complex set of communication, group-interaction, management, and work-orientation skills. The committee views these additional components as coming from two sources.

First, the impact of educational content in these areas is very important. For example, education for management of the engineering function (as distinct from MBA-style management) is notably lacking in most curricula. Essential nontechnical skills such as written and oral communication, planning, and technical project management (including management of the individual's own work and career) are not sufficiently emphasized. The question is, how to include training in these skills? The existing four-year curriculum is already severely strained, and the instruction-intensive nature of education for these skills makes teaching them even more problematical given the current high student-faculty ratio in schools. Five-year and dual-degree programs are two options; continuing education also holds promise. The committee believes that different schools can and should develop different means of accommodating these educational needs, depending on what each school deems important. Some will weave them into existing courses by changing the way in which courses are taught. Others will offer separate for-credit courses, using greater flexibility in course requirements. But some restructuring of the standard four-year curriculum will probably be required.

The second aspect of nontechnical education comes through work experience. The committee believes there is more educational value in early work experience than has generally been acknowledged. It imparts a greater ability to work in teams and a familiarity with project work. It gives invaluable experience in the use of equipment and instrumentation (severely curtailed in some schools by large classes and a lack of modern laboratory equipment). Most important, it sharpens the student's perspective on the relative importance of different aspects of the undergraduate education. The traditional sources of early work experience are cooperative education and summer employment. Cooperative education has some traditional problems: inconsistent support by industry, high program management costs to the schools, and faulty design of programs from the standpoint of industry are among those most often mentioned. But these problems are solvable. The committee recommends that academic and industry leaders join together with government as necessary to develop mechanisms for improving existing work-education approaches and devising new options to include a greater part of the engineering student cohort.

WHAT IS THE EMPLOYMENT PICTURE FOR ENGINEERS IN THE UNITED STATES?

In 1970, engineers represented 1. 6 percent of the U. S. work force. As of 1983 that figure was 1.4 percent. The percentage of engineers has dropped because of a rapid growth in the overall employed population; the number of engineers grows substantially each year—it is now approximately 1.6 million. Industry demand for engineers has been high for the past decade, notwithstanding the intervening recession. The perception of abundant jobs in engineering is reflected in the greatly increased enrollments in engineering schools. Demand has been particularly high in fast-growing industries such as electrical, electronics, and computer engineering. Spot shortages have appeared in these fields, but output from the engineering schools may by now be alleviating those shortages.

The committee found that, on average, engineers are the highest-paid professionals who are not self-employed. They enjoy among the lowest unemployment rates of any group (rarely higher than 2 percent). The most prevalent occupational areas axe development (28 percent) and general management (20 percent). The least frequent areas of work are research (less than 5 percent, with only 1 percent in basic research) and teaching (2 percent); however, 11 percent of all women engineers are involved in research.

One finding that was initially troubling was that there are apparently far fewer technicians and technologists[1] in the work force than there are engineers. This apparent weakness in engineering support seemed to imply inefficient use of resources. However, the committee found that self-reporting of data distorts the picture considerably (i.e., many technicians and most technologists define themselves as engineers). In addition, there are many engineers who do technician-level work. Thus, there is a built-in asymmetry in the data for these groups; the occupational structure is actually not as top heavy as it would appear. Regardless of the relative distribution of educational levels, the system seems to find the most appropriate balance via market mechanisms. Thus there is no need to redress the technician/technologist/engineer balance.

The data problem is further complicated by the fact that *engineering, engineers*, and the *engineering community* are poorly defined terms. Inconsistencies in definition pervade statistical studies, thus compounding the difficulty in understanding. Data bases and conceptual diagrams of the engineering community all reflect this lack of consistency. In the course of its work, therefore, the committee adopted comprehensive definitions of these terms.

Both directly and indirectly, the federal government has become a significant user of engineering goods and services. About 6 percent of engineers are employed directly by the government; a higher proportion of engineers work in the government (some 5 percent) than is found in either the industrial or academic sectors. When indirect employment is taken into account (i.e., prime contractors), the federal government employs some 30 percent of U.S. engineers. (It should be noted that this is roughly equivalent to the portion of the overall GNP generated by the federal government). The committee is concerned that civil service regulations make it difficult for the federal govern

---

[1] Technologists are defined as holders of a bachelor of engineering technology degree.

ment to compensate engineering employees at certain levels of experience (and in most engineering disciplines) in a competitive fashion. In view of the strong direct dependency on engineering talent for many of its most important activities, the federal government should review its compensation policies to ensure that it can competitively recruit and maintain a high-quality engineering work force.

There are serious concerns about the dislocation of engineers that takes place when major changes in demand occur. Often, shifts in government funding drive these changes. Although the profession as a whole has shown great adaptability to changing demand, such events cause considerable stress for individuals and within disciplines. Changes also result in inefficient use of engineering resources. Retraining programs offered by industry or government are of course one solution to this problem. However, the committee believes that effective continuing education throughout a career holds great promise for keeping engineers professionally flexible enough to anticipate and avoid harm from technological obsolescence and changing demand. The educational services offered by technical and professional engineering societies are important in this regard and should be supported and used by a greater proportion of the engineering community.

Although the committee did not look closely at the use of engineers from a managerial standpoint, many findings suggest that this is an important issue. The ways in which engineering resources and capabilities are allocated have an enormous bearing on the effectiveness of engineering practice in the United States. How an engineering enterprise is organized and managed can have considerable impact on productivity. Appropriate management practices can foster an atmosphere in which the creative, innovative potential of engineering is more fully tapped.

Thus there is a need for corporations and government agencies to examine critically the relationship between their engineering management practices and general management goals. Attention to these issues could have significant positive implications for the effectiveness of an organization.

ARE WOMEN AND MINORITIES ADEQUATELY REPRESENTED?

Since the early 1970s, considerable effort has been devoted to increasing the participation of women and minorities in engineering. The recruitment efforts have paid off: the percentage of minorities in the engineering work force has doubled and the percentage of women has more than tripled. Currently, more than 15 percent of engineering undergraduate students are women (as compared to about 1 percent in 1970), which has generated a feeling of success among many of those concerned with the issue.

However, some sobering facts should be pointed out. Compared with the sciences and other professional disciplines, women are still a small part of the engineering work force. Perhaps even more significant, beginning in 1982 there has been a mild slowdown in enrollments of women in engineering.

Similar trends can be seen for minority groups. Enrollments of blacks, Hispanics, and American Indians increased steadily throughout the 1970s but have recently leveled off or declined somewhat. The one exception to this pattern has been Asian Americans, who continue to study engineering at increasingly high rates. As in the case of women, minorities overall (with the exception, again, of Asian Americans) are poorly represented in the engineering work force in comparison with other professions.

What is the desirable level for these different groups? Some assert that it should be parity or near parity on a population-proportional basis. Women constitute about 50 percent of the general population and minorities constitute some 28 percent. Yet only 5.7 percent of engineers are women and 4.6 percent are minorities. On this basis, women are less well represented than the aggregate of minorities. However, Asian Americans alone account for nearly two-thirds of the total minority representation; blacks account for less than one-third. Because blacks constitute some 12 percent of the general population, it can be seen that on this basis their representation is roughly equivalent to that of women. The same pattern is reflected in the engineering schools, whether in comparison with the general population or with enrollments in other courses of study.

The committee believes that the determination of appropriate levels of representation in engineering for both women and minorities is not a matter for judgment by panels of educators and industry representatives. These are social questions requiring broader discussion. However, both women and minorities are represented as students and as practitioners in engineering at lower levels than in other science and technology professions. Therefore, the committee concludes that the participation of women and minorities in engineering should be matters of continuing concern to the engineering community. There is still much to be done.

A case in point is the treatment of women on engineering faculties. There is a recurring perception of bias against female faculty members in assignment of teaching responsibilities, in selection for research teams, and in granting tenure. In many schools there also appears to some to be a bias against female graduate students as candidates for faculty positions and in the provision of financial and intellectual support. College administrators should make a candid assessment of the attractiveness of academic life for women in their institutions, and if negative aspects such as these are found, they should take firm steps to eliminate them.

Another area needing attention is the precollege education of women and minorities in both science and mathematics. For women, early exposure to physics in particular appears to be a key factor in the later choice of engineering as a course of study. Poor preparation in science and mathematics limits the appeal of engineering to these groups and increases the attrition among those who do study engineering, especially among minority students. Educators should develop strategies to increase the size of the initial science/mathematics pool of minorities and to reduce attrition all along the educational pipeline. Such strategies should include innovative ways to increase the appeal of mathematics and physics for female students.

WILL THE ENGINEERING COMMUNITY, BE ABLE TO MEET FUTURE DEMAND?

Questions of supply and demand and of the relative balance between them have often occupied those concerned with engineering personnel resources. However, it is misleading to refer to an overall balance in supply and demand because the picture always varies considerably across different engineering disciplines. For example, demand for civil engineers is now less than the supply, while demand for computer engineers exceeds supply. The situation is always dynamic, although on average it may appear relatively stable. In fact, the difference between stringent shortage and painful surplus is a matter of only about 5 percent of the engineering pool in either direction.

There is little point in attempting to make projections of future shortages or surpluses of engineers. Demand cannot be predicted accurately. The committee does not know what economic turns the future will bring. The exact nature and timing of future technology development is also uncertain: New technologies will emerge, but no one can predict when or what they will be. International factors are also important. Will American companies increasingly go outside the United States for new business? Will foreign engineers increasingly compete with U.S. engineers for domestic as well as international business?

The best that can be done in the face of such uncertainty is to identify the changes that are likely to occur and then determine whether the system can cope with those changes.

The committee believes that there will be an increase in engineering work in the future. New technologies and the new industries they spawn will be at the center of this growth. Public expectations regarding health, safety, and environmental protection will also contribute, as will further development of third world countries.

At the same time, the productivity of engineers will also increase. This change will be based not just on increases in production and quality but on fundamental changes in the nature of engineering work brought about by new technologies and new engineering practices. Engineering tools based on the computer, such as computer-aided design and computer-based workstations, are part of this revolutionary change. New methods, such as simulation and modeling, are driving engineering activity in the direction of greater abstraction —more mathematical analysis and less experimentation.

The rate of change in each of these areas will vary from field to field, industry to industry. The degree of balance between the trends across different fields of engineering will have a major impact on the composition of the engineering community—on the ratio between engineers, technologists, and technicians and, indeed, on how we define engineering work.

Other factors will undoubtedly influence the scale and pattern of demand in different ways. Recurrent shortages of capital resources and shortages of both energy and raw materials will affect rates of growth in every field. Increases in the length of time over which industry seeks to maximize profits may ultimately result in improved product quality and thus in increased demand for technology-intensive goods. Government demand for engineering goods and services will probably increase even beyond present levels.

Underlying all these variables and uncertainties is as least one certainty: we are entering an era in which engineering will play a more dominant role than ever before. Requirements for both the quantity and quality of engineers are increasing.

The changes just outlined will have a great impact on how engineers are educated. Under such conditions, they will have to be adaptable as changing market and economic conditions force them to shift into new areas of work. Through better grounding in engineering fundamentals, more structured programs in continuing education, and greater preparation for managing engineering work and an engineering career, there may be a great increase in the self-directedness of engineers in general. Thus, in the future engineers may play a greater role not only in shaping the course of their own careers but also in determining the direction of development in engineering-intensive industries.

The engineering profession historically has demonstrated consider able flexibility and adaptability in responding to changing demand. This capability is likely to be taxed to the utmost in coming years. To meet the challenges the future will pose for engineering requires serious attention by government, industry, and academic leaders.

CONCLUSION AND PERSPECTIVE

When the National Science Foundation asked the National Research Council to conduct a study of the education and utilization of engineers, there were widespread concerns that the profession was under stress and that engineering education was in crisis. However, by 1984, during the period when this committee was conducting its phase of the study, data became available that suggested the situation might be improving. Engineering faculty were no longer leaving the schools at a significantly greater rate than they were coming in from industry. More students were beginning to pursue the doctoral degree, thus offering hope that faculty numbers might be augmented. Large numbers of students were responding to market demand, studying engineering and then going into industry. To be sure, this heavy enrollment created severe overcrowding in classrooms, but the graduates were largely bright, energetic, and ambitious and appeared to be satisfying industry's requirements.

Moreover, the engineering profession appeared to be healthy. It was no longer (at least for the moment) being subjected to the degree of criticism it had met with in the recent past. Engineers themselves are relatively well paid and enjoy the lowest overall unemployment rate of any occupation. It appeared to the committee that the engineering community was addressing many of its problems on its own. Market forces and the profession's traditional resiliency seemed to be having a salutary effect.

In reviewing these apparent trends, the committee then asked the questions, "Is action required, and, if so, what kind? Will the engineering enterprise in the United States retain its basic health in the absence of action?"

The committee concluded that inaction would pose risks that should not and need not be taken. Technological, economic, and social change will continue to intensify and will place even greater stresses on engineering's ability to adapt. Although some problems of the past appear to have been eased in recent years, whether the system will function well enough to meet the nation's needs in the future cannot be predicted or guaranteed.

Because the ability of the engineering community to meet society's changing demands in the context of a more competitive world is critical to the nation's interests, the committee believes that every precaution must be taken to ensure that it does function well.

Many areas continue to pose problems for engineering. Some require changes in funding; others require changes in current practice or simply changes in attitude. Some are relatively simple to implement; others are more difficult or complex. All are important. The consequences of ignoring the engineering enterprise are too great to permit the nation to take the future health of that enterprise for granted. Accordingly, the committee presents its recommendations for action.

RECOMMENDATIONS

It should be pointed out that these recommendations[2] do not derive directly from the foregoing executive summary, nor does the summary itself provide adequate support for the recommendations. Instead, the recommendations are drawn selectively from the accompanying report of the committee, which is itself based upon nine panel reports. In the executive summary, the committee has tried to distill the essence of this very complex set of reports and the extensive study that they represent. To gain a full understanding of the rationale upon which each recommendation is based, the reader is urged to read the report of the committee and to refer as well to the relevant panel reports.

1. Engineering institutions, such as industrial concerns and engineering schools, have proven in the past to be remarkably adaptable, and individual engineers generally have been flexible in responding to change caused by new programs and changing technology. The engineering system, although resilient, is not invulnerable; it requires proper financial, educational, and management support. *The committee concludes that there is no need for actions that would fundamentally alter the functioning of this adaptable system. However, there are serious problems of support, of curricula, and of policy and practice that must be addressed if that adaptability and flexibility are to be maintained.* (See chapter 5, pages 102–105.)
2. A shortage of highly qualified faculty continues to threaten the quality of engineering education. *Universities must take steps to make engineering faculty careers more attractive than at present in order to*

[2] Among the activities contemplated in a later phase (dissemination of results) of this study are presentations to representatives of industry, government, and academe and discussions of the recommendations of the study. From such interactions it is expected that additional initiatives and specific actions will be developed.

fill vacant faculty positions. Salaries need further improvement, adequate facilities are necessary, and current teaching overloads should be reduced. Such measures would help to alleviate the problem by increasing the number of highly qualified U.S. citizens who obtain the Ph.D. and choose teaching as a career. (See chapter 4, pages 53–56.)

3. A major increase in fellowship support and concomitant engineering college research support are needed in order to attract more of the very brightest U.S. citizens into graduate programs in engineering. *To attract top students into graduate work, doctoral fellowships should carry stipends equal to at least half the starting salary of a new B.S. graduate.* (See chapter 4, pages 56–59.)
4. *To assist in alleviating the faculty shortage, engineering faculty members and administrators should identify and utilize as faculty individuals such as government, military, and corporate retirees, with or without the Ph.D., who are not seeking tenure and who would welcome a short-term contract for a second career.* (See chapter 4, pages 66–68.)
5. If U.S. engineers are to be adequately prepared to meet future technological and competitive challenges, then the undergraduate engineering curriculum must emphasize broad engineering education, with strong grounding in fundamentals and science. In addition, the curriculum must be expanded to included greater exposure to a variety of nontechnical subjects (humanities, economics, sociology) as well as work orientational skills and knowledge. Education in these areas is needed to improve the communication skills of engineers as well as their ability to understand and adapt to changing conditions that affect technology development.

   To accomplish this expansion will require restructuring of the standard four-year curriculum by various means. *The committee recommends that extensive disciplinary specialization be postponed to the graduate level. Beyond that, individual engineering schools will have to closely examine their existing curriculum in order to ascertain how the curriculum can best be restructured to accommodate the other important educational needs.* (See chapter 4, pages 68–69 and chapter 5, pages 117–120.)
6. In the context of an increasingly global economy, American engineers must become more sensitive to cultural and regional differences, so that they can design products that foreign markets require and will accept. Engineers will also need to appreciate the financial, political, and security forces at play internationally. *The nontechnical components of engineering education ought to include exposure to these aspects of contemporary engineering.* In addition, the engineering com

munity should strive to ensure open communication on these matters among engineers and companies the world over. (See chapter 6, pages 114–115.)

7. The committee believes that cooperative education and other such interning programs have played a valuable role in undergraduate engineering education. *The committee therefore strongly recommends that the National Academy of Engineering and the professional societies take the initiative in bringing together representatives of industry, academe, and government to develop better work-study programs.* Means should be found to eliminate the sometimes cyclical nature of industry support for these programs and to make it feasible for a much larger fraction of the engineering student cohort to participate. (See chapter 4, pages 68–69.)
8. Patterns of government support since the 1950s have led to a two-tiered system of engineering colleges. As one result, colleges of the second tier (those that are primarily undergraduate-oriented) do not benefit sufficiently from the substantial government/ industry funding for graduate education and research at colleges of the first tier.

   The federal government and industry should recognize and support innovative programs in undergraduate engineering education in the second-tier institutions, which annually supply half of the nation's engineering graduates. These colleges must have access to new and additional sources of income. In addition, ways must be found to provide for more equitable distribution of the many benefits that accrue to first-tier schools. For example, faculty members and students at second-tier institutions will need to be involved in the use of research facilities and programs of major centers of research. (A plan for such access should be a part of the proposal for such facilities.) (See chapter 4, pages 61–63.)
9. With regard to the continuing problem of obsolete and deteriorating equipment and facilities in engineering schools, a national program of government-industry-college matching grants is required to address the situation. *Industry, academe, and the professional societies need to join forces in promoting legislation where necessary to facilitate gifts of laboratory equipment to colleges of engineering. In the special case of bricks and mortar, the federal government and industry should be prepared to match those funds raised for this purpose by state governments or from philanthropic sources.* (See chapter 4, page 60.)
10. Various organizations and institutions are developing programs (such as the Semiconductor Research Corporation and the National Science Foundations's Engineering Research Centers) designed to foster closer ties between engineering colleges and industry. *More such*

creative and innovative programs of a specific nature are needed to strengthen the bond between engineering schools and industry. Such initiatives ought to be in addition to current programs of industry support for shared faculty, advisory councils, and donations of equipment and funds. Continuation of the R&D tax credit is essential for maintaining all forms of industry funding of research in engineering schools. (See chapter 4, pages 76–78.)

11. The capacity of the engineering educational system could be expanded by creating a network of dual-degree programs such as those which already exist between some liberal arts and engineering colleges.

    The National Science Foundation should examine experience to date with dual-degree and other alternative programs, and should then take the initiative, if indicated, in establishing a pilot group of colleges and engineering schools to demonstrate effective structures for such programs. This pilot program could be funded by a combination of foundations, industry, and government agencies. Experience gained from the program could then be applied to a wider group of institutions. In addition, the experience gained would be relevant to the often-debated model of preprofessional followed by professional engineering education. It would also be highly relevant to the examination of options for restructuring the curriculum to satisfy competing educational demands (see recommendation 5). (See chapter 4, pages 66–68.)
12. Computers, and computer-aided instruction in particular, should be recognized as powerful educational systems tools. These tools should be applied as rapidly and as fully as practicable in all academic programs in such a way as to enhance the quality of engineering education. *Engineering schools should create programs for development of educational technology by faculty, with shared institutional, industry, and government funding.* (See chapter 4, page 71.)
13. Engineers can be productive in engineering work over a longer period if they have access to effective continuing education. However, the lack of company reimbursement and release time is a strong demotivator for pursuing continuing education. *Those companies that do not offer their engineering employees financial and worktime relief for continuing education are encouraged to do so.* (See chapter 4, pages 71–72.)
14. There is great variability among engineering technology programs in terms of entry requirements, standards of achievement, curricula content, semester hours required, and overall quality. The committee finds that this diversity serves a useful purpose, given the diversity of industrial needs in different regions. *However, technical and technology institutions should cooperate in eliminating variabil*

ity that has no relevance to market needs and is strictly arbitrary . (See chapter 4, pages 74–75.)

15. *To improve the qualifications of students intending to study engineering, it is essential to increase the number of high school graduates who are literate in science and mathematics; improved written and oral communication skills at the secondary level are also very important.* The committee supports the recommendations put forth in recent studies by the National Commission on Excellence in Education and by the National Science Board's Commission on Pre-College Education in Mathematics, Science and Technology. (See chapter 4, pages 73–74.)
16. Because of major demographic changes (such as a decline in the number of 18-year-olds and a population shift from the Frost Belt to the Sun Belt), schools in some geographical areas will experience significant decreases in application rates by the early 1990s. *Engineering schools should examine the impact of these factors in their area in order to anticipate steps they will need to take to increase the flow of qualified students from their regional pool.* One way to accomplish this is to increase the enrollment of qualified women and minorities. Other programs specific to the circumstances of the individual institution will also need to be devised. (See chapter 4, pages 62–66.)
17. While the fraction of women engineering students has grown considerably in recent years, it is still significantly lower than female representation in other fields of college study. Likewise, the proportion of women engineers is considerably lower than the proportion of women in other science/technology professions. *Therefore, continued efforts should be made to increase the participation of women in engineering*. Perhaps the most important elements are greater effort (as recommended by other study groups) to increase the study of mathematics and science by female secondary-school students and continuing action by colleges of engineering to increase female enrollment.

    It is also important to improve the role model represented by women engineering faculty. To this end, college administrators should make a candid assessment of the attractiveness of academic life for women on their faculties, and if negative aspects are found, they should take firm steps to eliminate them. (See chapter 4, pages 62–66 and chapter 5, pages 92–94.)
18. The committee recognizes the fine work being done in many cities and regions to encourage minorities to enter engineering school, as well as that of the many colleges and organizations which support retention programs for minority undergraduate engineering

students. Yet minorities continue to be underrepresented in engineering. *Therefore, the committee recommends that these efforts be broadened* . For example, precollege programs such as those operating in a few major cities and regions must be expanded and funded so as to better prepare and motivate minority students to pursue careers in engineering. (See chapter 4, pages 63–66.)

19. Existing definitions and diagrammatic conceptions of the engineering community are inconsistent and incomplete. Yet definitions and diagrams are essential as a basis for describing the engineering community and its essential elements in a manner conducive to accurate data collection, display, and analysis. *Therefore, the committee recommends that the National Academy of Engineering (NAE) take the initiative to call together the various public and private database-collecting organizations to see how best to arrive at commonality in definitions, survey methodology, and diagramming methodology*. Organizational roles can be determined in the coordinating meeting. The purpose will be to ensure, to the greatest degree possible, that data collection efforts result in accurate and compatible data bases that describe the engineering community and its various components in totality. (See chapter 3, pages 34–43.)
20. Data regarding engineering technologists and technicians indicate a top-heaviness in the work force, with engineers outnumbering these support personnel. However, this is a misleading impression deriving from asymmetry in the data. *Since the engineering occupational structure appears to find the most appropriate balance through market mechanisms, there is no need at the present time to take action to alter the technician/technologist/ engineer balance. However, periodic monitoring of this balance would be advisable*. (See chapter 5, pages 88–90.)
21. *In view of its strong direct dependency on engineering talent for many of its most important activities, the federal government should review its compensation policies to ensure that it can recruit competitively and maintain a high-quality engineering work force on a discipline-by-discipline basis*. (See chapter 5, pages 98–100.)
22. The committee believes that it would benefit the engineering community if a greater fraction of engineers were members of the engineering technical and professional societies. Therefore, steps should be taken to enhance the attractiveness of membership.

    Toward this end, the committee recommends that the activities of professional societies be explained more fully to students during the undergraduate years. In addition, industry and government agencies should encourage engineering employees to participate in the activi ties of the societies, and should provide support for that participation . (See chapter 3, pages 44–49.)

23. The engineering community has an obligation to assist the media in the media's job of informing the general public and various special constituencies regarding the nature and status of technical projects and programs. *To this end, the committee recommends that the NAE take the initiative in creating a media institute that would provide centralized coordination of a nationwide network of technological information sources to respond to media requests.* (See chapter 3, pages 44–49.)

BACKGROUND

Architectural drawing on overleaf from 1903, Municipal Archives, Department of Records and Information Services, City of New York

1

Introduction

CONTEXT OF THE STUDY

Engineering is a central feature of the technology development process. As such, it is a critical element in the economic fortunes of industrialized nations, both domestically and internationally. Our concern about the decline of smokestack industries in the United States and the growing strength of competitors such as Japan, as well as our enthusiasm for new products and new industries such as computers and biotechnology, are both linked to the health of engineering.

In recent years a rapid expansion in key disciplines of engineering has placed the profession under considerable stress. Calls of crisis have come from engineering schools, panels of business and professional leaders, and government reports. The highly publicized problems of our American economic system over the past several years, particularly in the face of increasing international economic competition, have helped focus even more attention on problems in the engineering field.

The greatest emphasis has been on problems in engineering education. [1] Faculty shortages, overcrowded classrooms, inadequate laboratory and teaching equipment, aging facilities, low graduate enroll

[1] One of the first tasks of the present study was to conduct a survey of recent publications on this subject and to compile a report of concerns and responses regarding engineering education. That report to the committee summarized the problems identified and solutions recommended by 66 separate reports, articles, and other documents.

merits, and questions about curriculum quality and content have been seen as central and pressing issues affecting many, though certainly not all, engineering colleges.

To a lesser extent, problems in the utilization of engineers in the contemporary American workplace have also been illuminated. Discussion has been directed at shortages of engineers in certain critical fields and surpluses in others. Another important concern is technological obsolescence and potential job displacement among engineers in rapidly changing fields. The rapid emergence of such new fields as semiconductor electronics and new engineering disciplines as bioengineering produces considerable stress on the system. Finally, the changing nature of engineering work has created confusion about what an engineer does and about the distinctions between engineers and other technical workers.

The many studies that have examined these problems in recent years have varied in identifying the key difficulties and their assessment of the urgency involved, although most have reported considerable cause for concern. A broad range of solutions has also been proposed, urging increased attention and assistance on the part of both industry and government as well as new attitudes and practices on the part of employers, universities, professional societies, and individual engineers. Finally, most of the previous studies have focused on only one side of the engineering equation—education or utilization, usually the former.

Problems relating to the social and sociological aspects of engineering have been given less emphasis in previous studies. There is, however, a persistent concern among engineers about their professional image and status in society. Related to this concern is the public's inadequate understanding of technological matters. The engineering profession has also encountered difficulties in satisfactorily attracting and incorporating blacks and other minorities. Questions about the recruitment and assimilation of women into engineering also persist.

THE COMMITTEE'S APPROACH

Because this critical time for the economy of the United States and the world seemed to call for synthesis, for examining the whole rather than the parts, the intention of the committee has been to look at problems of the engineering profession as an integrated whole, to devise workable solutions with potentially far-reaching effects, and to project those solutions in an effort to define and help structure the profession through the end of this century.

One of the most pressing requirements, especially for an integration effort such as this, was a means to identify and characterize the different elements of the engineering profession. Previous formulations and models were incomplete and did not take into account the ways in which the engineering profession adapts and responds to external and internal conditions. Accordingly, one of the primary activities of the committee was the effort to define contemporary engineering and the elements of the engineering community, to consolidate and analyze existing data pertaining to engineering, and to examine the dynamics of movement within and through the profession.

The committee also formed panels to examine in detail the different aspects of engineering and the institutional forces that influence the widespread and diverse engineering community. A subcommittee on engineering educational systems conducted a broad study of contemporary undergraduate and graduate engineering education, engineering technology education, and continuing education for engineers. A panel on engineering employment characteristics attempted to identify current patterns and trends in demographics and practice within the engineering community and to assess the capabilities of the engineering work force relative to present and future national needs. A panel on support organizations focused on mechanisms by which government at different levels, industry, schools, private practitioners, and society at large can provide positive support for improved functioning of the engineering system. Last, a panel on engineering's interactions with society examined the development of the profession in the United States and attempted to characterize its changing role vis-à-vis society in general.[2]

The rationale for the panel reports was to cast as wide a net as possible while maintaining the goal of synthesis and integration. The resulting in-depth, coordinated reports thus provide the raw material necessary for understanding the engineering profession in this age of technological, social, economic, and political change both at home and worldwide.

REPORT STRUCTURE

This report of the committee is a crystallization of those themes that emerged out of the broad study. It begins with a brief discussion of the

---

[2] See Appendix A for an organizational description of the committee and panels and Appendix B for a list of panel reports and background papers prepared for the committee.

role of engineers and engineering in building and shaping America and in maintaining America's economic power, world influence, and high standard of living. The report next examines the status of engineering today. Chapter 3 looks at the nature of the work, the organizational and occupational structure of the profession, and its support network. Chapter 4 assesses the strengths and weaknesses of contemporary engineering education. Chapter 5 discusses characteristics and trends of the engineering work force.

Chapter 6 attempts to specify world economic and technological features to which the engineering community must be able to adapt in the year 2000. The report concludes with a review of some of the educational and professional characteristics that the engineering community must acquire or maintain in order to adapt successfully.

Findings, conclusions, and recommendations of the committee pertinent to each chapter are presented at the end of that chapter.

2

The Role of Engineering in America

ROOTS OF THE PROFESSION

Engineering began in America with the building of forts, arsenals, and roads. But the young nation needed civilian as well as military projects, and when Thomas Jefferson founded West Point in 1802 he enjoined its cadets to form a corps of civil engineers. Until that time, there had been virtually no American-born civilian engineers. The directors of such large public works projects as canals and municipal water supply systems were European and brought with them their European training and European technology (Pursell, 1980). Engineering schools were slow to emerge because the demand for engineering skills was slow to develop. For almost the first half of the nineteenth century, only West Point and Rensselaer Polytechnic Institute trained engineers—primarily to work on expanding of the canal and railroad systems and on military projects.

Meanwhile, an indigenous talent for metalworking was being nurtured in machine shops through experimentation in the production of arms, agricultural tools, and other implements. By 1850 this expertise had become sophisticated enough to be considered mechanical engineering. The centerpiece of American machine technology emerged as a standardized system for production of parts called the American system of manufacturing. This technique, combined with a notable penchant for innovation and simple, elegant design, began to provide the United States with technological autonomy and to build the foundations of an independent economic strength. America was on its way to being the great success story of the Industrial Revolution.

The expansion of the country by rail, canal, and road combined with a rapid increase in population to produce a great market for available goods of all kinds, along with a need for efficient communications and transportation systems and for the training to build them. To meet these and other educational needs, the federal government began in 1862 to support higher education. Under the auspices of the Morrill Act it created the federally subsidized public land-grant college system, which gave great impetus to engineering education and made possible a more scientific approach to technical problems.

As a result, the profession began to diversify. Out of civil engineering grew mining and metallurgical engineering. Mechanical engineering became more specialized. And by the beginning of the twentieth century, a new emphasis on science in engineering had spawned electrical and then chemical engineering. Industrial engineering (initially a branch of mechanical engineering) developed to systematize further the manufacturing process, especially in the burgeoning automobile industry. Work roles also diversified: while military and independent consulting engineers had earlier been the most important, corporations now became the predominant force for technology development, and specialized assignments within a project team became the rule (Noble, 1977).

Wars stimulated the development of engineering in this country. Taking World Wars I and II together, government direction of research and development (R&D) for the war effort led to postwar booms in chemical, aeronautical (later aerospace), radio, electronics, nuclear, and computer engineering. Even the Great Depression spurred engineering development through massive government funding of such projects as the Tennessee Valley Authority and the Rural Electrification Administration. Engineering had become the nucleus of the nation's phenomenal productivity and economic health. It underlay the rapid growth in such strong industries as steel, automobiles, agriculture, and manufacturing. It was a source of strength in good times and a source of salvation in times of duress.

By the end of World War II, the United States had the world's preeminent economy. Its political dominance, especially in the West, was inseparable from its economic dominance. The Marshall Plan was only the first installment of a global postwar strategy of using America's great wealth to provide aid to other nations, thus stimulating and then serving world markets as it built bonds of friendship and obligation. The foundation of America's postwar economic strength was based on innovation, productivity growth, and great economic scale —all dependent on engineering.

MAINTAINING AMERICAN STRENGTH AND INFLUENCE

Engineering played an indispensable role in establishing the United States' position in the world. That preeminence has been challenged before, and is being sorely tested now. At such times, the focus shifts to the role of engineering in maintaining U.S. power and influence. The profession has a critical role in maintaining the nation's defensive capability, a role that becomes more demanding with the increasing emphasis on technology in modern weapons. In addition, engineering must help maintain a thriving domestic economy. This requirement becomes more challenging as the service sector grows and as the U.S. share of international markets shrinks.

Other nations have rebounded from the devastation of World War II and are now confronting the United States with serious economic competition. Even the developing countries are seeing tremendous growth in the manufacture of goods and supply of services, including those of increasing technological sophistication. Thus the traditional importance of engineering in maintaining American strategic and defensive strength has come to be matched by its crucial role in maintaining U.S. economic competitiveness in the international marketplace. Both responsibilities depend on the problem-solving approach that is at the heart of engineering.

IMPROVING THE QUALITY OF LIFE

Engineering is also responsible to a great extent for enhancing the quality of life in the United States. It is no exaggeration to say that the profession has the same impact an the nation's social and economic health as the medical profession has on its physical health. This means, for one thing, helping to create and maintain the many systems necessary to support our large and affluent population. Highways and bridges, ground transportation systems, air transportation and traffic control, telephone and power utilities, water treatment and distribution, and waste treatment and management all form an extraordinarily complex network of facilities and services that are taken for granted for the most part and in which efficiency, safety, reliability, and low cost are expected by the public as a matter of routine.

At the same time, engineering provides the technical means by which government and industry are able to protect national resources and ensure public safety and the quality of life. This involves participating in the industrial regulatory process and developing the means to do so through testing, standards development, and so forth. Thus engineering is integrally involved in both producing economic growth and moderating its potential for harm. This simple fact provides one key to understanding the complexity of the engineering profession's role in modern America.

MAINTAINING THE PUBLIC TRUST

Another key to that complexity is the changing attitudes toward engineers, engineering, and technology in general. Along with the enormous increase in engineering activity in the postwar era has come an increase in the awareness and critical scrutiny of that activity by the public. Especially since the early 1960s, antitechnology attitudes have become prevalent as public attention has focused on the growing capacity of technology for doing harm to individuals, the environment, and society itself. There have been many different concerns—the environmental and health effects of air and water pollution, problems of safety in the design of automobiles and other products, the use of technology in the Vietnam War, and fears about nuclear power, among others. But all of them led to an atmosphere of mistrust regarding the objectives of technology development and the basic morality of its purveyor, the engineer (Report of the Panel on Engineering Interactions with Society).

Since the mid-1970s the public attitude seems to have swung in the other direction. Views of engineering and technology are now for the most part once again positive (see, for example, Yankelovich, 1984). However, the times of naive public acceptance of the wonders of modern technology are now forever past. The public is better educated than ever before, and its current enthusiasm for technology development is probably not permanent. The residue of the antitechnology attitude means that engineers have new social responsibilities added to their traditional technical responsibilities. They must continue to improve the quality of life and spur the economy through the goods and services they produce, while at the same time anticipating and avoiding adverse social consequences of their work. This is a considerable challenge, and the actions needed to see that it continues to be met over coming decades are at the heart of this report.

References

Noble, D. F. American by Design: Science, Technology, and the Rise of Corporate Capitalism. New York: Alfred A. Knopf, 1977.

Pursell, C. W., Jr., ed. Technology in America: A History of Individuals and Ideas. Cambridge, Mass.: MIT Press, 1981.

Report of the Panel on Engineering Interactions with Society, in preparation.

Yankelovich, D. Science and the public process: Why the gap must close. Issues in Science and Technology (Fall 1984), pp. 6–12.

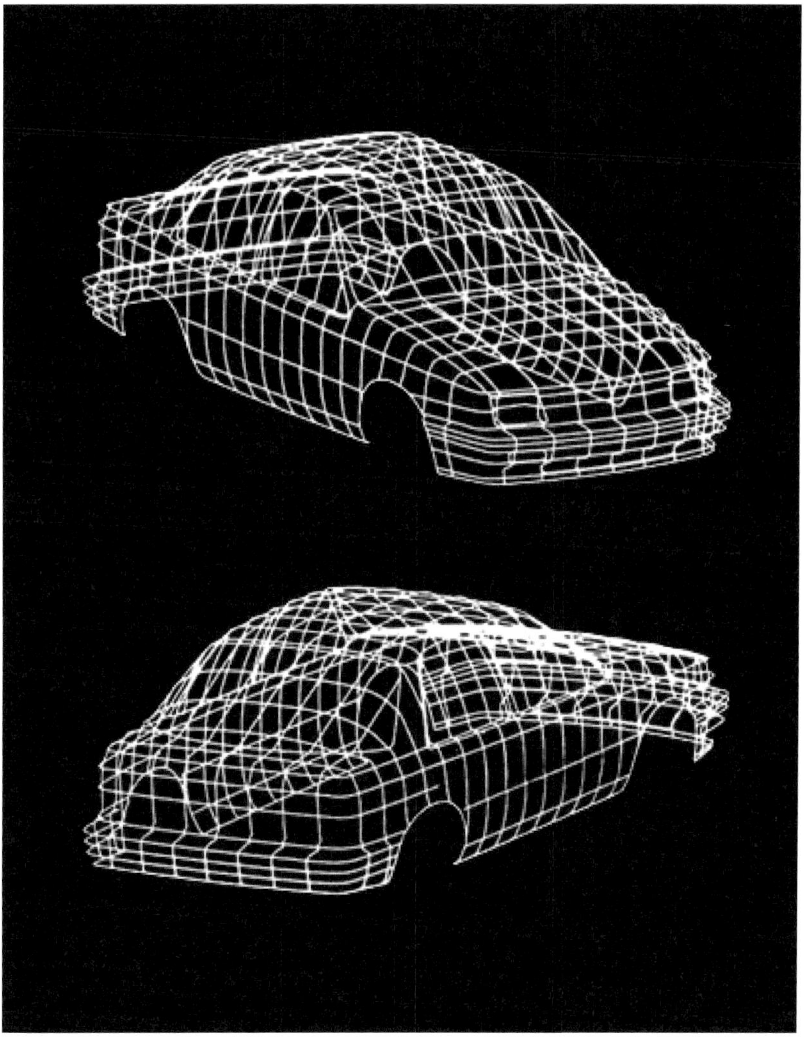

WHERE DOES ENGINEERING STAND IN AMERICA TODAY?

Photograph on overleaf courtesy of the Chrysler Corporation

3

Defining the Engineering Community

THE CHANGING NATURE OF CONTEMPORARY ENGINEERING

Once the roles of engineers were fairly uniform and clearcut. On one hand military and civil engineers performed marvels in taming the continent and opening it up for development. On the other hand inventors and early corporate engineers helped build U.S. industrial might, providing materials and machines to make life easier and more pleasurable. From this era derive the romantic images of the engineer: the lone surveyor in boots and Mackinaw, the wizard inventor in his workshop, and the master of the industrial dynamo. These images embodied the heroic concept of the engineer celebrated in American folklore until very recently.

Throughout this century, however, the nature of engineering work has been changing steadily. The corporate engineer has come to predominate, with work characterized by large project teams, relative individual anonymity, and dedication to discrete bits of technology advancement in a highly specialized field. Business itself has changed. Modern corporations are generally much larger, more bureaucratic than their earlier counterparts, and dominated by professional managers, often with little technical background. The global sphere of operations of the modern corporation is another new factor (Report of the Panel on Engineering Interactions with Society.) Rapid technological change has also led to the greater diversification of engineering disciplines. Today there are numerous engineering specialties, each repre sented by a professional society and each composed of highly defined subdisciplines that may eventually emerge as separate disciplines in their own right.[1]

Compounding the trend toward diversification has been the emergence of entire new fields of engineering because of scientific breakthroughs or clusters of technological breakthroughs that offer the potential for completely new products. Examples would certainly include the invention of the transistor in 1948, which inaugurated the field of solid state electronics. Some 10 years later, the development of the integrated circuit in turn produced rapid expansion in computer science and engineering.

Such developments have had an enormous effect on engineers, on engineering schools, on engineering-intensive industries, and on the way engineering work is done. For instance, not only have many engineers become involved in the research, development, and design of computers and computer systems, but a new tool has made possible better designs in less time. This same advance has also given rise to a new support person—the computer specialist. By 1982 there were more than 500,000 of these professionals (Bureau of Labor Statistics, 1983). Today, the emergence of the very large scale integrated circuit is about to create another revolution in the computer field as it sets the stage for artificial intelligence and other potential "fifth-generation" applications.

Another emerging field is that branch of bioengineering called biochemical engineering (for sometimes biotechnology). Built on scientific breakthroughs in genetics and molecular biology, it might seem an unlikely candidate as the basis for extensive engineering activity, but the number of companies operating in this field has grown rapidly since the late 1970s, and the number and usefulness of potential products are seemingly endless (Office of Technology Assessment, 1984a).

Some of these emerging technologies will profoundly alter the nature and practice of engineering, particularly if several gain momentum at the same time. Automation in manufacturing, including computer-aided design and manufacturing, and computer-aided engineering, are examples. Composite materials and artificial intelligence are progressing rapidly and are undergoing intensive R&D. Yet the nature of technology development is such that technologies that initially appear

[1] Some of the major currently recognized engineering disciplines are civil, mining, mechanical, electrical and electronic, computer, chemical, aeronautical/aerospace, manufacturing, industrial, petroleum, marine, agricultural, nuclear, bioengineering, engineering mechanics, environmental, and ceramic, metallurgical, and materials.

promising sometimes do not come to fruition and are abandoned, while others emerge unexpectedly. Ferroelectric technology is an example of the former; random access disk memory is an example of the latter.

The impact of a new technology such as the computer is usually pervasive throughout engineering. Not only does it change the way engineering work is done in every discipline, but it can also change the amount of engineering work available. For example, the full impact of manufacturing automation cannot yet be predicted with confidence. However, it is certain that it will affect engineers as well as craftworkers (Office of Technology Assessment, 1984b).

In addition to new fields and specialties of engineering, there has also been a diversification of engineering-style activity within the corporate framework. For example, technologists (usually holders of a bachelor's degree in engineering technology) and technicians (frequently holders of associate degrees in engineering technology) today often perform work that formerly required engineers; they are frequently qualified for many specialized or routine engineering tasks, and this is therefore the most efficient use of human resources.

Conversely, engineers are often found to be doing repetitive, routine tasks rather than the creative tasks for which they are educated. The reasons for this seemingly inefficient use of resources may include regulations, automation, poor management, administrative necessity, or even individual preference. In addition, engineers often work in management, sales, and other support positions. The complexity of the overall picture thus contributes to a prevailing confusion about what engineering is and what an engineer does.

Another factor in the changing nature of engineering—and this is characteristic of many occupations today—is that engineers enter and leave the labor pool at different times and for different reasons. An individual may enter the work force after graduation, leave it to return to school for advanced study, then return to practice in a different specialization; he or she may leave industry for a teaching position, or vice versa, or may leave temporarily to raise a family. Major shifts in demand for engineers in a certain field may bring large numbers back into engineering from other occupations or, conversely, may cause large numbers of practitioners to leave engineering entirely. Engineers also leave the field for better opportunities, to pursue other interests, or to retire. Foreign-born engineers may be required by the conditions of their visas to leave the country. This constant flux in the engineering work force makes it more difficult to characterize accurately the engineering profession, to determine with any certainty who is what, how many engineers there are, and where they work.

CHARACTERIZING ENGINEERING'S INFRASTRUCTURE

All these factors of change have caused a blurring of the concept of engineering within our society. Yet a clear understanding of the profession is necessary as a basis for national policymaking, for fiscal and economic planning, and in general for gaining a better understanding of how the technology development process works—crucial knowledge in today's world. Consequently, the Panel on Infrastructure Diagramming and Modeling undertook the task of developing information and tools that could improve our understanding of the engineering profession in the contemporary context.

Objectives and Accomplishments

The panel asked of what components the engineering profession consists and how it functions as a system. To that end the panel formulated a set of definitions relating to the concepts "engineer" and "engineering." In addition, it developed a set of flow diagrams that provide, at varying levels of detail, a representational basis for understanding and quantifying the dynamics of the engineering system.

The committee believes that the results of this effort represent a major contribution toward achieving those goals. They have withstood sustained scrutiny and criticism on the part of the panel members, the committee as a whole, and a range of interested outside observers. Although the definitions and diagrams may need modification to accord with the specific needs of future efforts in this direction, they nevertheless provide a firm and rational base upon which future contributions can be built.

After developing the representational flow diagrams, the panel next attempted to fill in the comprehensive diagram with data for different years. In the process it found that the existing data bases, although numerous and extensive, are inadequate for that purpose, for reasons described later. Thus the panel was able to flesh out the diagram only partially and with considerable uncertainty. Future efforts along these lines will have to begin with the standardization and consolidation of data bases relating to engineering personnel resources.

In addition to the foregoing, the panel conjectured that a computer model could be developed that would represent the overall flow diagram and thus make it more useful. Such an investigative tool could provide a controlled "what if" capability for evaluating assumptions within a low-cost study environment. However, an attempt to develop a model sufficiently detailed and comprehensive to analyze the flows described in the overall diagram would have required resources beyond those available to the panel or the committee.[2] Therefore, the panel decided to develop an interim simulation model as an aid in analyzing flows for the purposes of its study. The resulting model represents a small subset of the overall diagram. It is limited in its capabilities, and is not predictive with any degree of reliability. However, the panel concluded on the basis of its experience that such models could be useful for gaining insights and drawing broad conclusions about cause-and-effect relationships.

The following sections describe in greater detail the activities and findings associated with these development efforts.

Definitions

As a starting point, the panel saw the need to define engineering in the broadest possible way so as to include all those activities that constitute the engineering function. The panel then developed the concept of an engineering community consisting not just of degreed engineers but of all those involved in engineering work, support of engineering work, or engineering education, whether they be engineers, scientists, technologists, or technicians. This all-encompassing approach provides a "universe" that was deemed necessary for describing adequately the complex dynamics seen within engineering practice today and anticipated for the future. The definitions follow:

- **Engineering**[3]**.** Business, government, academic, or individual efforts in which knowledge of mathematical and/or natural[4] sciences is employed in research, development, design, manufacturing, systems engineering, or technical operations with the objective of creating and/or delivering systems, products, processes, and/or services of a technical nature and content intended for use.

[2] The National Science Foundation is developing a model that will be capable of serving that purpose in the long term (National Research Council, 1984.)

[3] The precise wording of the definition of engineering produced considerable controversy within the panel and the committee. Debate focused on whether the phrase "intended for use" was strong enough to convey the basic motivation or intention of engineering. Many of the panel members felt strongly that the definition should convey the notion of optimization or economy in the design and delivery of the engineering product. Yet such qualifiers (however true or desirable) also seemed to make the definition more complex and diffuse—perhaps unnecessarily so, because the intention "for use" could be construed to imply optimization directed at the needs of the user.

[4] Including physical sciences.

- **Engineering Community.** People meeting at least one of the following conditions:

  a. Actively engaged in engineering, as defined previously.
  b. Actively engaged in engineering education.
  c. Qualified as an engineer, engineering technologist, or engineering technician (see definitions below) and actively engaged in such engineering support functions as engineering management or administration, technical sales, or technical product purchasing.
  d. Qualified as an engineer, engineering technologist, or engineering technician and was but is not now actively engaged in engineering, engineering education, or engineering support. An important point is that the definition of engineering—as well as the other definitions developed by the panel—is by no means an idealized one. It is not meant to prescribe or judge. It is designed to facilitate the collection and analysis of data about the engineering community. Other definitions (for example, those used by the Accreditation Board for Engineering and Technology) are qualified so that they focus on different aspects of the engineering function; those definitions are thus appropriate for the particular purposes for which they were formulated.

The panel's next step was to define the members of the engineering community. The definitions of engineer, engineering technologist, and engineering technician were set forth in terms that were specific but also inclusive, again so as not to place artificial restrictions on attempts to model the real world of engineering. The occupational definitions are:

- **Engineer.** A person having at least one of the following qualifications:

  a. College/university B.S. or advanced degree in an accredited engineering program.
  b. Membership in a recognized engineering society at a professional level.
  c. Registered or licensed as an engineer by a governmental agency.
  d. Current or recent employment in a job classification requiring engineering work at a professional level.

- **Engineering Technologist.** A person having at least one of the following qualifications:

  a. A bachelor's degree from an accredited program in engineering technology.
  b. Current or recent employment in engineering work, but not qualified as an engineer as defined above.

- **Engineering Technician.** A person having at least one of the following qualifications:

    a. A degree or certificate from a one-to-three-year accredited technical program.
    b. Current or recent employment in engineering work, but not qualified as an engineer as defined above and at a lower job level than that required of an engineering technologist.

While the occupational definitions differ little from those employed in previous studies and reports, the notion of an engineering community that is far broader than a mere community of engineers is a distinct departure from most earlier approaches. The panel's initial examination of flows of personnel into and out of activities that are decidedly engineering made it clear that individuals without formal education in engineering would have to be taken into account, as would all those not currently engaged in engineering work but nevertheless qualified by virtue of training or experience to become active as the need might arise. In addition, technical personnel engaged in engineering support functions would have to be included. To leave out any of these categories of people would, it was felt, greatly oversimplify the description of the way that engineering work is performed and engineering needs are met today.

Flow Diagrams

As described earlier, in order to provide a representational basis for understanding and quantifying the dynamics of this complex system defined as the engineering community, the panel developed a series of flow diagrams. The basic flow diagram (Figure 1) provides a simple representation of the flows of people into engineering education and employment, and their eventual exit from the engineering community. The comprehensive flow diagram (Figure 2.) is an expansion of the basic diagram. It depicts all the significant sources, flows, and activities of different elements of the community.

In Figure 2, aggregated pools of people (defined as "stocks") are identified that make up the sources from the population at large, the various modes of educational preparation for entry into the engineering community, categories and flows of people actively engaged in engineering-related work, the technical reserve of potentially active participants, and the various modes of permanent exit from engineering.

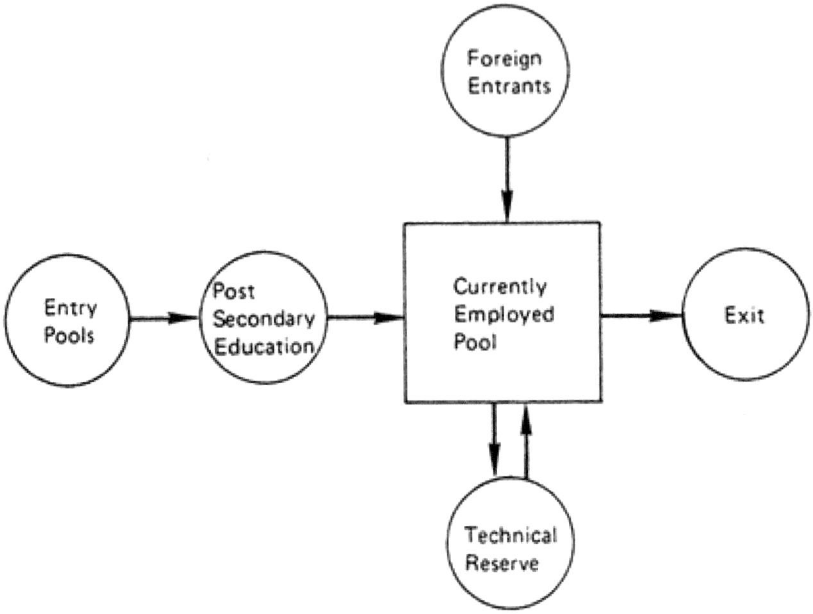

Figure 1
Basic flow diagram for the U.S. engineering community.

Based on the comprehensive diagram, the panel next elaborated a series of detailed flow diagrams, each of which focuses on one particular "stock" and tracks all flows into and out of that pool. At this level of detail, it becomes practical to associate numbers of people with the discrete pools and flows. A variety of existing data bases (see the next section) were employed to quantify individual data elements within each detailed diagram, thereby giving a series of one-year snapshots of flows into and out of that particular pool. Figure 3 is 1 of 19 such detailed diagrams; the example shown here depicts the flows of U.S. secondary school students. The alphanumeric codes in parentheses are data-element labels. Table 1 presents the corresponding numeric values for each data element in the diagram in three different years. Of course, these illustrations are not easily interpreted without a fuller explanation of their meaning. For more information, see the Report of the Panel on Infrastructure Diagramming and Modeling.

Apart from the obvious value in having a graphic representation of a complex system, the availability of these flow diagrams affords a number of important benefits:

- It reduces the ambiguity involved in dealing with technical human resources by establishing a consistent, clearly defined set of relationships among the groups involved.
- It provides a framework for use in quantifying the various pools and flows.
- It permits the tracking of past events with respect to the engineering community and can be used as a basis or framework for forecasting future problems.
- It provides a standardized model for studying the behavior of subsets of the community.

As a case in point, developing the comprehensive flow diagram led to the identification of two large populations not fully recognized previously—a technical reserve pool and a staff support pool. These are essential and integral elements of the engineering community; without them the functioning of that community cannot thoroughly be understood. In addition, the development effort revealed that the engineering community has a greater complexity in structure and flows than has generally been appreciated.

Data Bases

Fourteen or more data bases, considered significant, were used to obtain data and estimates on the education and employment of groups making up the engineering community. These data bases had been compiled by a variety of national organizations and agencies concerned with technical personnel.[5] While an enormous amount of information was available, a number of difficulties were encountered in using the existing data bases to derive values for the flow diagrams.

One problem was a lack of compatibility among data bases because of the diversity of purposes for which they had been compiled. Lack of consistency in the definitions used by various compilers was also a problem. Because of the differing needs of data base managers, there are differences in the focus of data bases (for example, how scientists and engineers are employed versus where they are employed). As a result, there are marked differences in measurement criteria from one data

[5] The main data base sources were the Engineering Manpower Commission (EMC) and the Bureau of the Census—primarily data collectors only—and the National Science Foundation, the National Research Council, the Bureau of Labor Statistics, and the National Center for Education Statistics—all data collectors as well as interpreters.

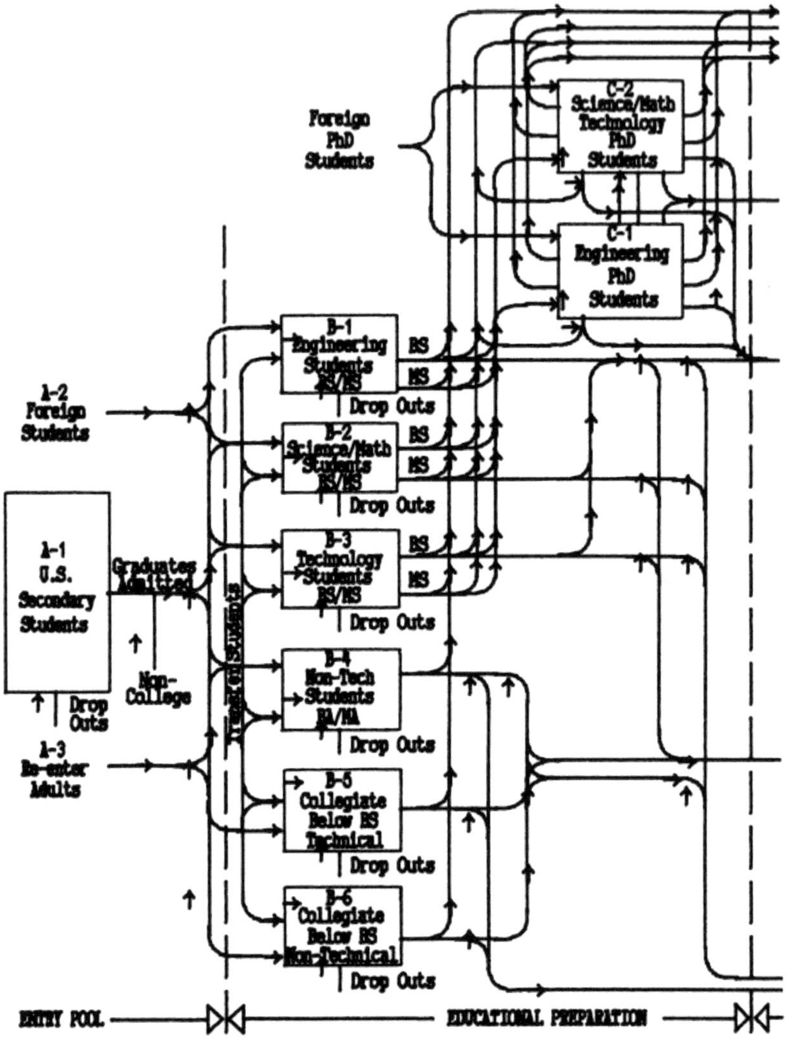

Figure 2
Comprehensive flow diagram for the U.S. engineering community.

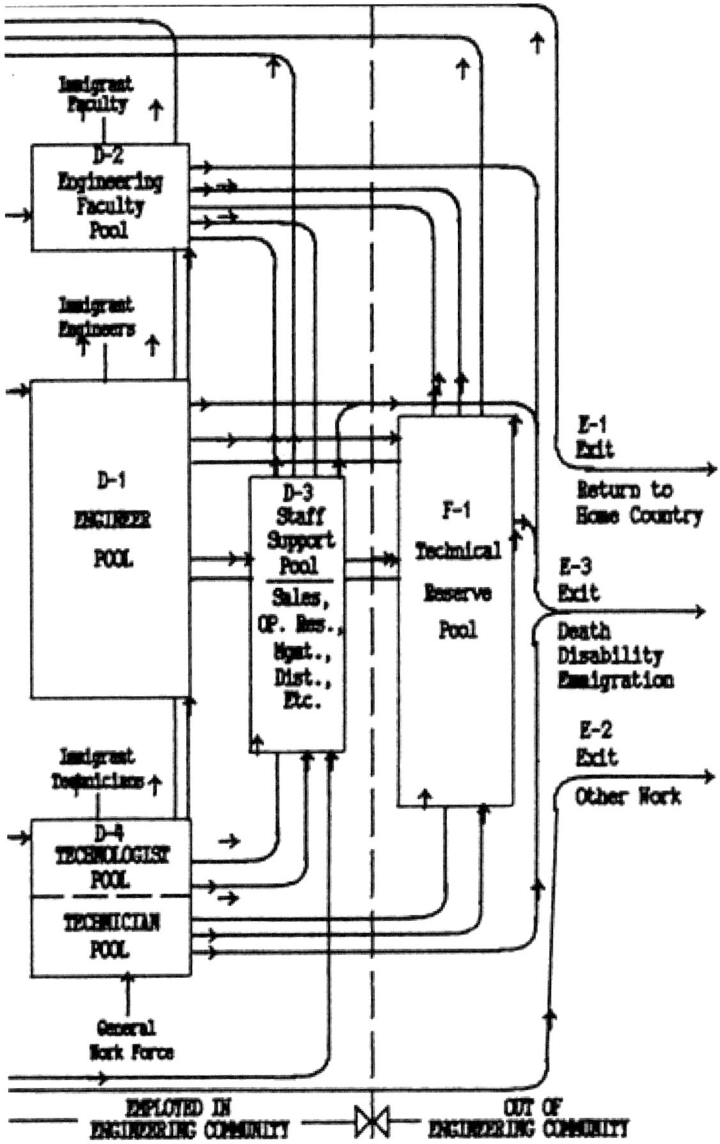
Immigrant Faculty
D-2 Engineering Faculty Pool
Immigrant Engineers
D-1 ENGINEER POOL
D-3 Staff Support Pool
Sales, OP. Res., Mgmt., Dist., Etc.
F-1 Technical Reserve Pool
E-1 Exit
Return to Home Country
E-3 Exit
Death Disability Emigration
E-2 Exit
Other Work
Immigrant Technicians
D-4 TECHNOLOGIST POOL
TECHNICIAN POOL
General Work Force
EMPLOYED IN ENGINEERING COMMUNITY
OUT OF ENGINEERING COMMUNITY

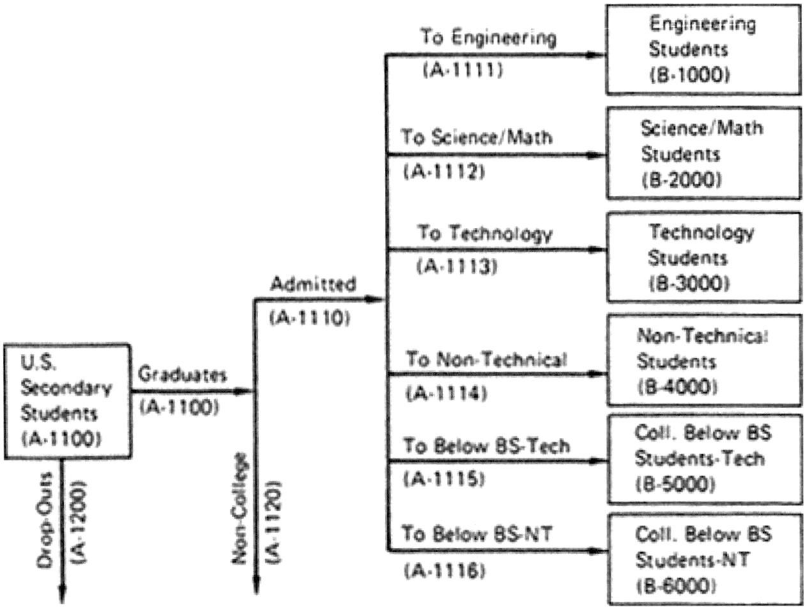

Figure 3
Flows of U.S. secondary students (A-1).

TABLE 1 Numerical Data on Flows of U.S. Secondary Students (Thousands)a

Label	Description	1960	1970	1960
A-1000	U.S. Secondary Students	9,600.0	14,418.0	15,191,0
A-1100	Secondary School Graduates	1,864.0	2,896.0	3,063.0
A-1110	Admitted to College	929.8	2,080.2	2,625.1
A-1111	To Engineering	67.6	71.7	110.1
A-1112	Science/Math[b]	92.3	134.9	143.7
A-1113	Technology		4.8	11.0
A-1114	Nontechnical 4 years plus	460.1	801.6	944.6
A-1115	Collegiate Below BS-T	20.2	60.0	150.7
A-1116	Collegiate Below BS-NT	718.8	363.5	748.3
A-1120	Noncollege			
A-1130	Nondegree College			
A-1200	High School Dropouts	998.0	929.0	1,099.0

[a] Includes foreign students

[b] Science/Math includes: agricultural/natural resources, biology, computer science, math, physical sciences, general science programs

base to another. There are also differences in choice of respondent (for example, individuals or households or establishments) and in the frequency of updating (varying from 1 month to 10 years). These differences result in significant discrepancies in personnel estimates.

Shortcomings of the individual data bases from the standpoint of the flow diagrams presented another problem. Overall, for example, the data bases fail to provide current information on nondegree or associate-degree engineers and computer specialists. A significant number of engineers are not degree holders or are upgrades; lack of such information is particularly important with regard to technicians, few of whom hold a B.S. degree. Coverage of gender, racial and ethnic background, citizenship, and income is uneven across the various data bases. There are limited data on the flows of students between engineering and other courses of study or across engineering disciplines. Additionally, the data bases often fail to distinguish among master's and doctoral students or to specify their disciplines. Data on the mobility of students between two-and four-year colleges are also lacking. These shortcomings are at least partly a function of the prevailing narrow definition of the engineering community. While they could be compensated for to some extent, the net impact on the flow diagrams developed by the panel is that data elements tend to underestimate the size of stocks and flows.

The unavailability of comprehensive, compatible data bases is made more disturbing by the fact that important data are not being used. An example is the Higher Education General Information Survey data, which are collected and filed by each state but not subjected to subsequent analysis until copies of the raw data are received by the National Center for Education Statistics. These data could be put to more immediate and fruitful use at minimal cost to the federal government if they were digitized at the state level (perhaps with federal funding.).

In short, currently available data bases provide only a limited understanding of the engineering community. Existing data were inadequate for making historical comparisons or for constructing consistent portraits of the engineering community, past or present.

To rectify this serious problem, the committee recommends that the National Academy of Engineering take the initiative to call together the various public and private data-collecting organizations to see how best to arrive at common definitions, survey methodologies, and diagramming methodologies. The purpose of this coordination will be to ensure to the greatest degree possible that data collection efforts result in accurate and compatible data bases that describe the engineering community and its various components in totality.

The Engineering Personnel Model

As described earlier, the Panel on Infrastructure Diagramming and Modeling developed and tested a simple computer-based model of the dynamics characterizing the engineering community. The model developed by the panel is neither econometric nor predictive, that is, it cannot take into account the impact of such external and unpredictable factors as a change in defense spending or a recession. The model merely provides a snapshot of a selected flowpath in which a change in a parameter at one end of the path produces a corresponding change in a parameter at the other end.

A very restricted set of objectives was chosen for the model. Using the terms of the comprehensive flow diagram, the flowpath selected was that for population to education to job market (Figure 4). The model was limited to a relatively high summary level, and only people with degrees in engineering, physical science, mathematics, and computer science were included. Finally, the model was run in an open loop mode to permit easier interaction and model formulation.

The resulting model can simulate the flow of engineers in the United States beginning in 1950. It is also possible to run alternative cases, thus affording a relatively crude form of forecast, but because of the restrictions set, the model cannot offer projections at a high level of confidence. (Thirty-year projections based on the statistics of 1950, for example, give results that deviate from the actual by as much as two to one.) Indeed, the panel concluded that the rate and unpredictability of change in technology and in the economic, political, and social spheres precludes the development of any reliable predictive model. However, such a model can offer constructive guidance for future causal studies and for the development of new educational policies.[6]

For a further discussion of modeling and its potential, see the Report of the Panel on Infrastructure Diagramming and Modeling.

THE SUPPORT STRUCTURE FOR ENGINEERING

Along with characterizing of the engineering community, it is important to understand the organizational context within and through which that community functions. Accordingly, the panel on support

---

[6] Detailed data on the model and its structure and example runs are available on request from the Office of Scientific and Engineering Personnel of the National Research Council.

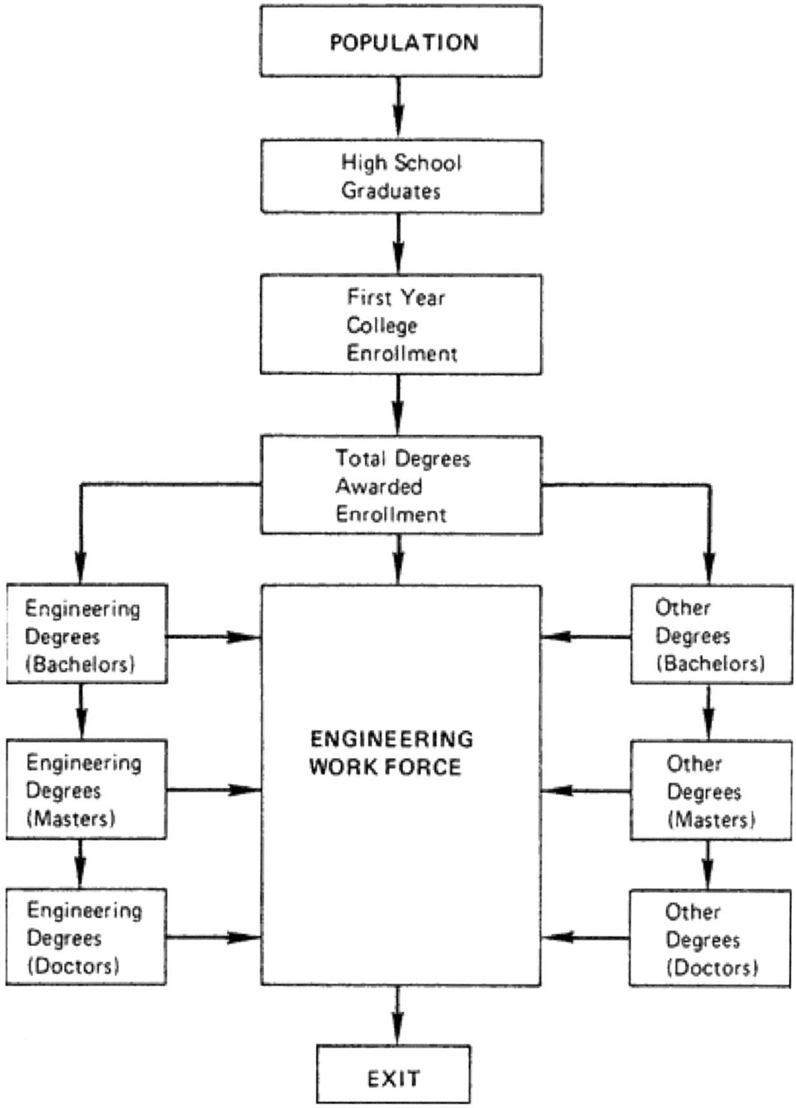

Figure 4
Engineering personnel model diagram.

organizations undertook to identify the types of supporting organizations and the needs they serve with respect to various sectors of the engineering community. These sectors include engineering academic institutions, government, industry, and private practice. In addition, the needs of society at large were considered in terms of its ability to affect the engineering profession in both positive and adverse ways.

The panel identified a wide variety of needs that exist in each of the four sectors of the engineering community and which must be met by specific types of support organizations. One finding was that common support needs exist across the four sectors. There are common needs for:

- Maintaining technical competence through continuing education.
- Information exchange and ready access to essential information.
- Continued professional development, defined in terms of the profession as a whole and in terms of individual development.
- Professional standards and ethics, involving on the one hand assurance that engineering is functioning responsibly and on the other a greater understanding by society of the effort of the engineering community to be responsive to its needs.

A great many organizations serve these and other needs of engineers, including, very prominently, the employing organizations and society itself (through government agencies, legislatures, and schools). Engineering educational institutions serve the needs of all sectors, not only through their primary mission of providing degree-oriented instruction but also in offering continuing educational opportunities, conducting research, and facilitating information exchange on many levels.

The panel also took special note of the role played by engineering associations and societies in support of engineers and engineering. There are over 50 such societies and associations (and more than 400 if state and local organizations are counted). They fall into five major groups:

- Those focused primarily on established or emerging engineering disciplines, such as the American Society of Civil Engineers and the Institute of Electrical and Electronic Engineers.
- Those focused on practice in a broad occupational field, such as the Society of Automotive Engineers, the American Institute of Aeronautics and Astronautics, and the American Society for Engineering Education.

- Those focused on a specific technology or group of technologies, such as the American Nuclear Society, the American Society of Safety Engineers, and the Society of Manufacturing Engineers.
- Those formed to promote and serve the professional and nontechnical interests of their members, such as the National Society of Professional Engineers, the Society of Black Engineers, the Society of Hispanic Engineers, and the Society of Women Engineers.
- Those societies formed by consortia of other societies to accomplish different and sometimes complementary profession-wide missions. Examples of these are the American Association of Engineering Societies, the National Action Council for Minorities in Engineering, and the Accreditation Board for Engineering and Technology. Standards-setting organizations such as the American National Standards Institute and the American Society for Testing and Materials are also in this category.

These voluntary organizations provide an extremely wide and varied range of support functions, including publishing technical information and general professional news, presenting seminars and symposia, offering guidance and scholarships to students, representing the interests of engineering in public policy forums, and providing public information about engineers and engineering achievements. A very valuable area of activity is the setting of standards, including technical standards, professional standards of conduct, and engineering educational standards.

The technical/professional societies have generally been very effective in meeting the needs of the engineering community—in particular those of their members. However, there is a concern that too small a proportion of the engineering community actually supports these efforts. Taking into account overlapping memberships, the panel made a rough estimate that perhaps only about one-third of the total engineering work force actually belongs to one or more societies, although many more enjoy occasional or indirect benefits from the diverse support services that these societies provide.

The committee believes that it would benefit the engineering community if a greater fraction of engineers were members of the engineering technical and professional societies. Therefore steps should be taken to enhance the attractiveness of membership.

Toward this end the committee recommends that the activities of professional societies be explained more fully to students during the undergraduate years. In addition, industry and government agencies should encourage engineering employees to participate in the activities of the societies and should provide support for that participation. Greater industry support should come as a result of more aggressive efforts on the part of professional societies to make industry management aware of the many benefits provided by the societies to industry management as well as to the individual engineer.

One area in which the technical/professional societies have had only limited success—notwithstanding much effort over the years—is in communicating the nature and value of the engineering endeavor to society at large. Many of the problems that engineers must face—career dislocation through sudden shifts in demand, problems of professional image and ethics, proliferating regulatory legislation, and inadequate funding for engineering education, for example—are related to a poor comprehension (and even apprehension) on the part of the general public about the engineering community and its works.

The engineering community to some extent has contributed to this isolation of itself from the public by an attitude of elitism and by a reluctance to discuss often-controversial and complex (as well as proprietary) technical matters with the media. Often, too, engineers mistrust the motives of reporters. Yet these attitudes, however understandable, are self-defeating. Today, the public's perception of technology has become a major factor affecting the country's decision-making processes. Public attitudes toward engineering in general have become more positive in recent years, but there is no guarantee of permanency in this trend.

Since the general public depends on the mass media for most of its information, any effort to improve public understanding of engineering must focus on improving media coverage. The engineering community must actively help the media in this regard. Mechanisms for improving media coverage are for the most part already in place, and need only be strengthened and expanded. The various support organizations—engineering professional societies, government agencies, and engineering schools and corporations—can broaden their existing public information programs vis-à-vis the public and the media. In addition, existing science and technology media services can be used more fully to provide an effective interface between engineers and the media. Organizations of the latter type exist (for example, the Media Resource Service of the Scientists' Institute for Public Information, in New York City), but are not universally used.

The committee recommend that the National Academy of Engineering take the initiative to create a media institute that would provide centralized coordination of a nationwide network of technological information sources. This institute would explicitly not be a public relations organization. It would not initiate contacts with the media; rather, it would respond to media requests for information. Funding for this network should come from four sources: the government, through the National Academy of Engineering; media organizations; engineering societies; and corporations. This four-part funding could be useful for ensuring the public credibility that such a network must maintain if it is to succeed. The committee is convinced that, properly implemented, this approach would be effective in improving public understanding of engineering and the engineering enterprise.

FINDINGS, CONCLUSIONS, AND RECOMMENDATIONS

1a. Comprehensive, objective definitions are essential as a basis for describing the engineering community and its constituent elements in general terms. Such definitions are also indispensable for accurate collection, display, and analysis of data about the profession.

1b. To understand adequately the flows and relationships of groupings within the engineering community, the panel found it necessary to construct a comprehensive flow diagram. Development of the flow diagram led to the identification of two large populations—a technical reserve pool and a staff support pool—that are essential and integral elements of the engineering community.

1c. Currently available data bases provide only a limited understanding of the engineering community. Existing data were found to be inadequate for making historical comparisons or constructing consistent portraits of the engineering community. There is a strong need for a more comprehensive and consistent set of data, available on an annual basis, for use in tracking and assessing the supply and utilization of engineers.

The committee recommends that the National Academy of Engineering take the initiative to call together the various public and private data-base-collecting organizations to see how best to arrive at commonality in definitions, survey methodology, and diagramming methodology. Organizational roles can be determined in the coordinating meeting. The purpose will be to ensure to the greatest degree possible that data collection efforts result in accurate and compatible data bases that describe the engineering community and its various components in totality.

2. The technical and professional societies have been generally very effective in meeting the needs of the engineering community—in

particular, those of their members. Although the activities and products of the societies are available to all, only an estimated one-third of the engineering community supports these societies with their money and talent.

Steps should be taken to enhance the attractiveness of membership in the technical and professional societies. Toward this end, the committee recommends that the activities of the societies be explained more fully to students during the undergraduate years. In addition, industry and government agencies should encourage engineering employees to participate in the activities of the societies and should provide support for that participation.

3. Many of the problems that engineers must face are related to a poor comprehension (and even apprehension) on the part of the general public about the engineering community and its works. Yet today, the public's perception of technology is a major factor affecting our country's decision-making processes. Because the general public depends on the mass media for most of its information, any effort to improve public understanding of engineering must focus on improving media coverage.

The committee recommends that the NAE take the initiative in creating a "media institute" that would provide centralized coordination of a nationwide network of technological information sources to respond to media requests for information.

References

Bureau of Labor Statistics. 1983. Occupation Employment Survey, 1983.

Fechter, A. 1984. The talent pool for engineering: Are the colleges the only source? Paper presented at the Joint Meeting of the Scientific Manpower Commission and the Engineering Manpower Commission.

National Research Council. 1984. Labor Market Conditions for Engineers: Is There a Shortage? Proceedings of a Symposium held by the Office of Scientific and Engineering Personnel. Washington, D.C.: National Academy Press.

Office of Technology Assessment. 1984a. Computerized Manufacturing Automation: Employment, Education, and the Workplace (OTA-CIT-235). Washington, D.C.: U.S. Government Printing Office.

Office of Technology Assessment. 1984b. Commercial Biotechnology: An International Analysis [OTA-BA-218]. Washington, D.C.: U.S. Government Printing Office.

Report of the Panel on Engineering Interactions with Society, in preparation.

Report of the Panel on Infrastructure Diagramming and Modeling, in preparation.

Report of the Panel on Support Organizations and the Engineering Community, in preparation.

4

Current Status of Engineering Education

As was pointed out in the introduction, the most critical and concerned attention directed at the engineering profession in recent years has focused on engineering education. This is where the cries of crisis have been most frequent and insistent. The educational system is correctly perceived as producing not just the fodder of the technology development process, but its seed corn as well. The training, skills, and knowledge of recent graduates are of critical importance to that development process, and trends that threaten their continued supply to any degree also threaten the foundations of industry and the national economy.

The linkage between engineering innovation, quality, and productivity on the one hand, and industrial and economic strength on the other is clearly evident as we look around at the world today. That linkage is two-directional in nature; occurrences with major economic impact also affect engineering. Figure 5 illustrates how closely the enrollment of engineers (and degrees awarded) in the United States is tied to national economic events, as well as to sociological attitudes (Report of the Panel on Infrastructure Diagramming and Modeling.). (It should be noted that underlying factors such as demographic shifts also affect the amplitude of these curves, as in number 10 on the chart.)

A primary objective of the committee was to reexamine the status of engineering education today, to see whether time and a degree of high-level attention to these problems in recent years might have brought about significant improvements in the situation.

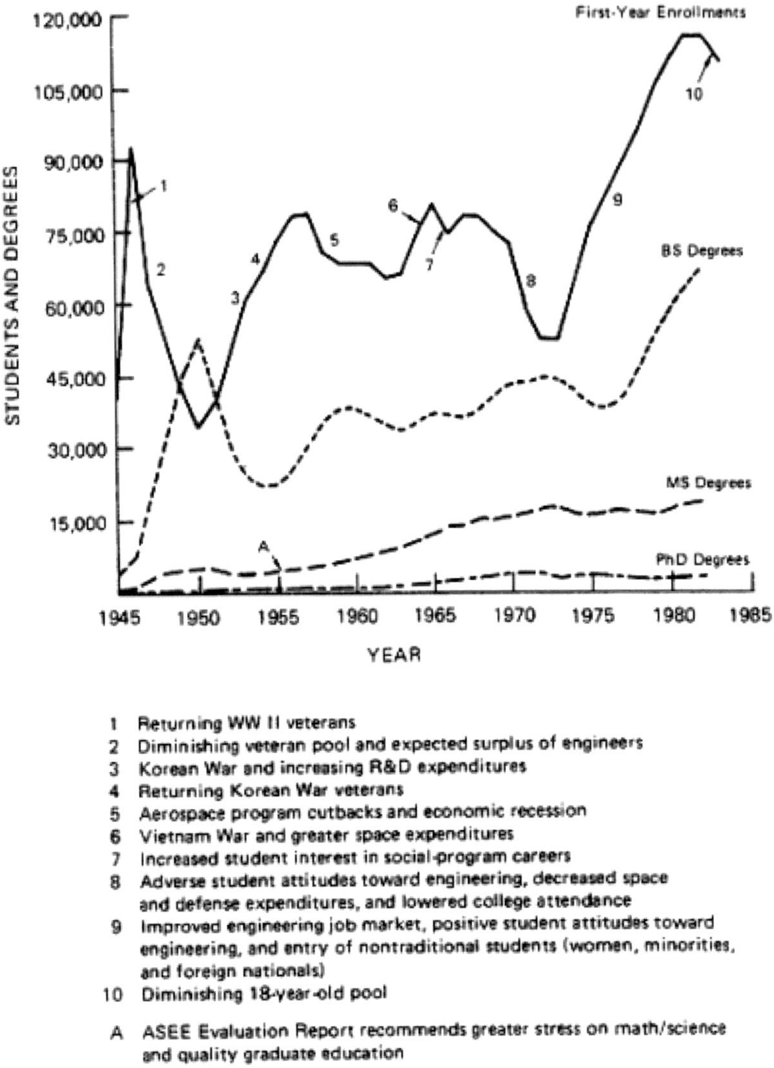

Figure 5
Engineering degrees and 1st-year enrollments: Historical factors influencing changes in engineering enrollments.

Four separate panels of the Subcommittee on Engineering Educational Systems examined relevant aspects of undergraduate education, graduate education and research, engineering technology education, and continuing education for engineers.

Based on the findings of those panels, it is possible to examine engineering education issues in a way that cuts across the different levels and types of programs. A useful organizing principle might be to look first at areas that are of critical importance—either because of their potential for doing harm or because of the timeliness of the needs they impose—and then to discuss special topics that are of broad or long-term importance. Finally, we will examine a number of points at which the educational system is experiencing significant change.

CRITICAL AREAS

Faculty

If there is one immediately pressing problem in engineering education, it is the current shortage of engineering faculty. Estimates of the severity of the shortage range from 1,567 to 6,700 (1,567 is the number of unfilled positions reported in a survey of engineering deans in 1983, and 6,700 is the number necessary to restore the student/faculty ratio to the levels of 1967–1969 and 1975–1976; see the Report of the Panel on Graduate Education and Research). The most recent survey of engineering colleges conducted by the American Society for Engineering Education (ASEE) revealed that 8.5 percent of budgeted faculty positions were unfilled in the fall of 1983 (American Society for Engineering Education, 1984b). Data derived from long-term analysis of advertisements for faculty positions indicate that 8.5 percent is higher than normal. The committee roughly estimates that the norm is probably around 3 or 4 percent.

The lack of sufficient faculty is the most important factor currently limiting attempts to increase the quality, scope, and number of engineering programs.

The shortage has several contributory causes, including the perceived unattractiveness of a teaching career relative to a career in industry and a decrease in available Ph.D.s in combination with a rapid increase in student enrollments in recent years. The latter has resulted in overcrowded classrooms that are themselves a further disincentive to teaching: student/faculty ratios rose 37 percent between 1976 and 1982 (Report of the Panel on Graduate Education and Research). A major concern has been that these ratios are too high and that they reduce the student-faculty interaction that is essential to high-quality education. Also frequently cited as negative aspects of a teaching career are noncompetitive salaries and poor research facilities compared with those available in industry.

In order to attack the faculty shortage problem, the ASEE Engineering Deans' Council recently adopted the following policy statement to encourage top-quality students to consider careers as engineering faculty members (American Society for Engineering Education, 1984a):

> At least 1000 intelligent and highly motivated individuals with doctoral degrees in engineering will be needed every year as faculty members in institutions of higher learning in the United States. Charged with the critical responsibility of educating prospective engineers, these individuals must enjoy the challenges and satisfaction of teaching, the excitement of research at the very frontiers of knowledge, and the freedom of self-direction. The opportunities for a lifelong, productive, satisfying, and rewarding career are unlimited.

Some have argued that engineering schools should be able to handle increased student loads through increased productivity of existing faculty with no loss of educational quality. Greater use of teaching assistants is one conventional approach for reducing a professor's per-class workload. But teaching assistants require money in the form of graduate assistantships, and such funds have perennially been in short supply.

In addition to the current shortfall of faculty, there is a continuing need to replace retiring faculty members. Because of present age distributions among engineering faculty, it can be expected that some 7,000 will retire over the next 15 years—an average of about 450 per year, probably increasing from 300 per year in the near term to 600 per year by the turn of the century (Report of the Panel on Graduate Education and Research).

The matter of low academic salaries has also been perceived as a major disincentive, and the perception has undoubtedly steered many young potential faculty members away from teaching. Although there are signs of improvement in this regard at some schools, there is still a considerable disparity between academic and industry salaries (Engineering Manpower Commission, 1983c, 1983d.).

There are several points that should be made here that complicate salary comparisons. First, faculty salaries at every level must be compared with those of Ph.D. engineers in industry. In addition, an equitable comparison for full professors is with industry supervisory Ph.D. holders (division heads), because some full professors are recruited into these positions. However, academic salaries are for 9 months. Because many faculty (including nearly all entry-level faculty) obtain research grants for 2 summer months, salary comparisons should reflect that augmentation (i.e., a multiplier of 11/9 must be applied).[1] Figure 6 compares adjusted salaries of Ph.D.-holders employed in industry and academe.

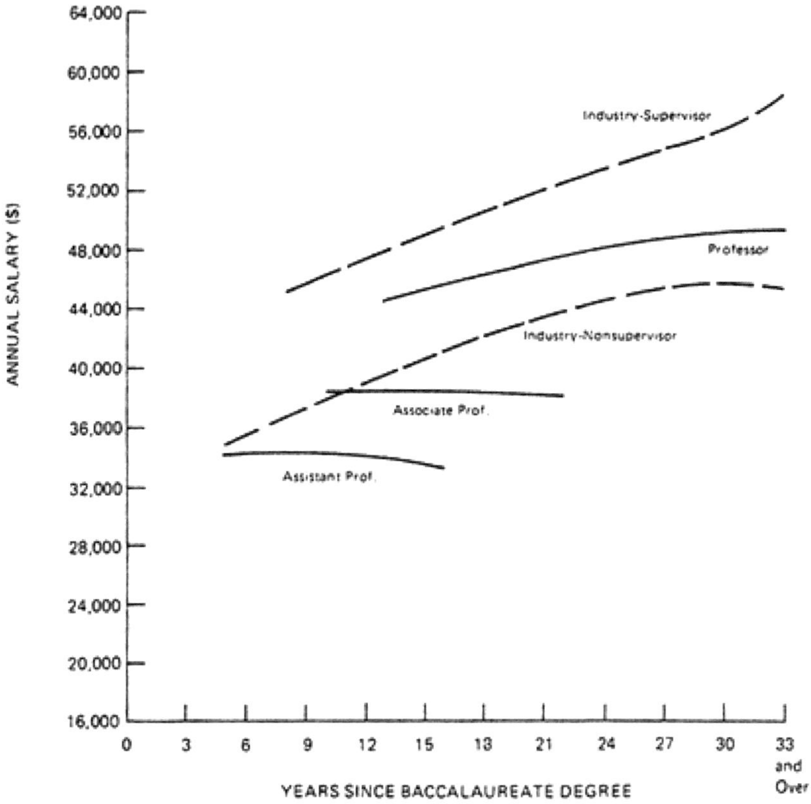

Figure 6
Comparison of academe-industry engineering Ph.D. salaries (all professional salaries adjusted to 11-month basis). Source: Engineering Manpower Commission, AAES, 1983.

Even these adjusted industrial-academic comparisons may be deceptive, however, because they involve median salaries. This approach ignores (in the case of faculty) large school to school differences and many individual differences. For example, the salaries of some estab

[1] At some schools, funding for all 3 summer months is the case (requiring a multiplier of 12/9).

lished professors are substantially augmented by income from consulting or book royalties. Younger faculty generally do not have the time or opportunity to obtain these supplements to income. A crucial point is that for tenure-track positions schools typically attempt to hire the best doctoral engineers available. These same people can sometimes command significantly higher than median salaries in industry (as high as $45,000 to start, in some cases), so that the real disparity may be even greater than the chart indicates.

When all these factors are taken into account, the salary problem is a real one. The salaries of full professors are well below those of their counterparts in industry. Moreover, the key salary problem is with junior faculty—assistant and associate professors beyond the entry level—and this is of course what discourages many young Ph.D.s considering teaching as a career.

Graduate Degrees

Figure 5, at the beginning of this chapter, demonstrated that graduate degrees awarded have not kept pace with B.S. degrees in recent years. Doctoral degree output has been particularly hard hit. While the number of engineering bachelor's degrees increased by 81 percent between 1977 and 1983, full-time doctoral enrollment increased only 33 percent in the same period. Table 2 presents numerical B.S. /M.S. /Ph.D. comparisons. Note that total annual Ph.D. production has been roughly stable at about 2,800 in recent years, although it rose to about 3,000 in 1983 (Report of the Panel on Graduate Education and Research). Certainly a major reason for the lack of interest has been the starting salaries offered to B.S. engineers by industry, which are very attractive in comparison to the extremely low income afforded by graduate study.

However, this situation now appears to be changing. Based on current numbers of doctoral-level graduate students, the Panel on Graduate Education and Research projects that Ph.D. output will increase to approximately 4,000 in 1988 (see Figure 7). The question must now be asked: Will this increase solve the faculty shortage?

The Panel on Graduate Study and Research initially calculated that 3,900 engineering Ph.D.s per year would be required to meet the needs for faculty, given that industry demand does not increase substantially. However, the committee concludes that advancing technology will cause industry demand for engineering Ph.D.s to increase steadily throughout the coming years. In addition, about 40 percent of the Ph.D.s graduating in recent years have been foreign nationals on temporary visas. Therefore, the projected supply of 4,000 Ph.D.s per year

TABLE 2 U.S. Engineering Degrees 1950–1983

	Bachelor's Degrees		Master's Degrees		Doctor's Degrees	
Year Ending	Foreign Nationals[a]	Total	Foreign Nationals[a]	Total	Foreign Nationals[a]	Total
1950	n/a	48,160	n/a	4,865	n/a	492
1951	n/a	37,887	n/a	5,134	n/a	586
1952	n/a	27,155	n/a	4,132	n/a	586
1953	n/a	24,165	n/a	3,636	n/a	592
1954	n/a	22,236	n/a	4,078	n/a	590
1955	n/a	22,589	n/a	4,379	n/a	599
1956	n/a	26,306	n/a	4,589	n/a	610
1957	n/a	31,221	n/a	5,093	n/a	596
1958	n/a	35,332	n/a	5,669	n/a	647
1959	n/a	38,134	n/a	6,615	n/a	714
1960	n/a	37,808	n/a	6,989	n/a	786
1961	n/a	35,860	n/a	7,977	n/a	943
1962	n/a	34,735	n/a	8,909	n/a	1,207
1963	n/a	33,458	n/a	9,460	n/a	1,378
1964	n/a	35,226	n/a	10,927	n/a	1,693
1965	n/a	36,691	n/a	12,246	n/a	2,124
1966	n/a	35,815	n/a	13,677	n/a	2,303
1967	n/a	36,186	n/a	13,887	n/a	2,614
1968	n/a	38,002	n/a	15,152	n/a	2,933
1969	n/a	39,972	n/a	14,980	n/a	3,387
1970	n/a	42,966	n/a	15,548	n/a	3,620
1971	1,565	43,167	2,930	16,383	741	3,640
1972	1,944	44,190	2,973	17,356	773	3,774
1973	2,136	43,429	2,551	17,152	708	3,587
1974	2,436	41,407	3,099	15,885	1,014	3,362
1975	2,468	38,210	3,250	15,773	891	3,138
1976	2,799	37,970	3,628	16,506	1,060	2,977
1977	2,996	40,095	3,825	16,551	993	2,813
1978	3,084	46,091	3,579	15,736	874	2,573
1979	3,788	52,598	3,944	15,624	929	2,815
1980	4,895	58,742	4,402	16,941	982	2,751
1981	5,622	62,935	4,589	17,643	1,054	2,841
1982	5,410	66,990	5,216	18,289	1,167	2,887
1983	6,151	72,471	5,145	19,673	1,179	3,023

Data from 1950–1952 taken from *Facilities and Opportunities for Graduate Study in Engineering*, American Society for Engineering Education, Washington, D.C., March 1968. Data from 1953–1976 supplied by Engineering Manpower Commission, New York, N.Y. Data for 1977–1979 from *Engineering Manpower Bulletin #50*, November 1979, Engineers Joint Council, New York, N.Y. 1980–1983 data from Engineering Manpower Commission.

[a] For these data, "foreign nationals" refers to non-U.S. citizens on temporary visas.

will be inadequate to meet the nation's needs—in particular, those of academia.

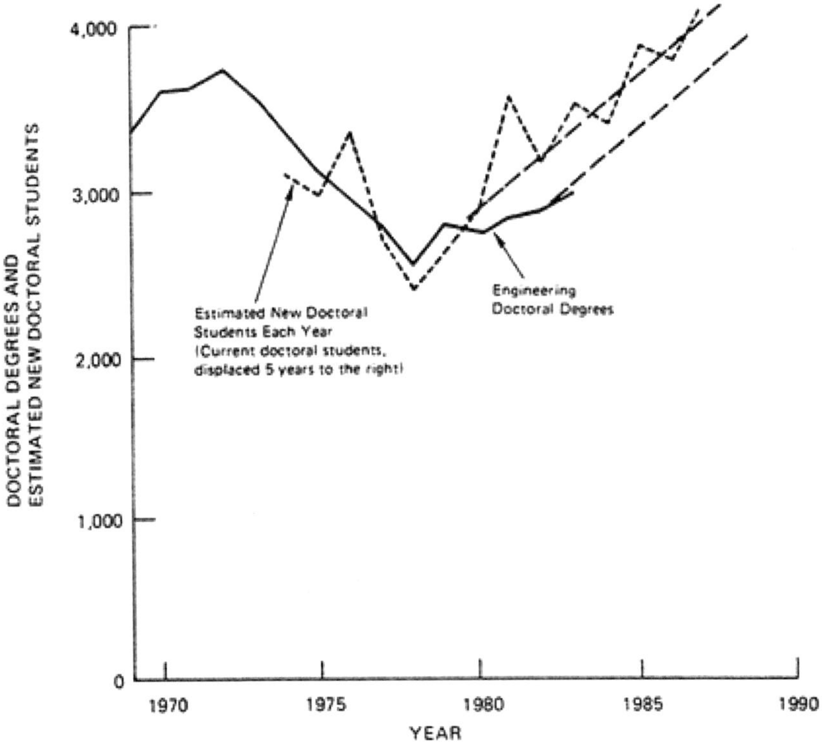

Figure 7
Engineering doctoral degrees per year. Source: Report of the Panel on Graduate Education and Research.

The percentage of Ph.D. students who are foreign nationals in the United States on temporary visas rose from about 14 percent in 1970 to about 42 percent in 1983 (Report of the Panel on Graduate Education and Research). Generally, only about half of these individuals expect to remain in the United States. Although foreign-born graduates of U.S. doctoral programs tend to go disproportionately into teaching (and in that sense have been the salvation of engineering education in recent years), the increase in the percentage of those who cannot stay in the United States threatens to dilute the advantage gained through increased Ph.D. output. On this basis, the committee concludes that the pool of doctoral candidates should include a higher proportion of U.S. residents.

To ensure that even the projected increase in Ph.D. output does in fact occur—that is, that it does not short-circuit into a large exodus at the master's level—and to increase the proportion and numbers of United States residents, will require additional funding by government and industry. The committee concludes that in order to minimize the financial disincentive, doctoral fellowships should carry stipends equal to at least half of the starting salary of a new B.S. graduate (or about $13,000 in 1984 dollars). Based on projected requirements for permanent-resident engineering Ph.D.s, the Panel on Graduate Education and Research estimates that 1,000 doctoral new starts per year will be needed. The panel calculated that these fellowships will cost the nation in the range of $60–$70 million per year, divided between the federal government and industry[2] (Report of the Panel on Graduate Education and Research).

Such figures can be misleading, however, in that they do not reflect a range of other costs that are driven by Ph.D. output and faculty growth. First, the additional doctoral production will require a corresponding increase in funded research. Second, more faculty will require more office space. Third, to improve the percentage of graduates opting for an academic career, careful attention must also be paid to starting faculty salaries. Until the Ph.D. offers a reasonable return on the investment of time, energy, and lost income, there will not be sufficient incentive for seeking it. Some universities are already addressing the latter problem.

Although the most serious concerns have focused on Ph.D. output, the importance of the master's degree should not be overlooked. In some areas of civil engineering and in most fields of electronics and computers, the M.S. has become the standard level of academic preparation for those engaged in design work. However, the proportion of M.S. degrees to B.S. degrees has been decreasing since about 1976 (Table 2).

The master's affords a level of specialization and familiarity with research practices not usually found in the B.S. graduate. Industry utilization of M.S. holders varies from company to company, from assignment to the same tasks as B.S. graduates to a more specialized role closer to the research-oriented work of Ph.D.s. Thus, this degree offers a versatility that is becoming increasingly important in light of the multidisciplinary and complex nature of much engineering work today.

[2] Assuming a 4-year Ph.D. program, with some attrition occurring, for a total of 3,500 students by the fourth year of the program; slight yearly increases in stipend, for an overall average of $14,000 per year per student; and an accompanying grant to the institution of $6,000 for tuition and fees.

Equipment Obsolescence

A major problem, alluded to earlier, is the age of teaching and research equipment in engineering colleges. One retired executive of a large U.S. corporation recently reported that, upon visiting his alma mater, he found engineering students in the laboratory using the same equipment he had used in the 1930s. The useful life span of laboratory equipment is currently considered to be about 10 years. The impact of new, advanced technologies and the rapidity of technological change are probably shortening that span even further. Yet the average age of laboratory equipment in engineering schools nationwide is 20 to 30 years (National Society of Professional Engineers, 1982).

Governmental and industrial support programs in this area have been sporadic, so that a serious mismatch exists between the need for equipment and the level of support. Obviously, the cost of state-of-the-art equipment is enormous. Even industry has substantial difficulty in remaining current. Yet the median age of instruments in the schools is about twice that of industry instrumentation. This means that industry gifts of used equipment to schools, while generous, are of limited value in increasing the technological currency of students and faculty. Leadership in engineering research in many fields has now clearly passed from schools to industry, so that the direction of technology transfer has reversed its traditional flow to a certain degree. Thus, this problem has major implications for the quality of education and the efficiency of the technology development process overall.

A related and important problem is seen in the aging of physical plants, including "bricks and mortar," in engineering schools. This condition is worsening at a time when the importance of engineering education to regional and national economic development is being recognized. For some time, the practice has been for the federal government and industry not to provide support for bricks and mortar. The committee urges a change in this practice.

A national program of government-industry-college matching grants is required to address the problem of equipment and facilities, including bricks and mortar. The federal government and industry should be prepared to match funds raised by colleges from state governments or from philanthropic sources for this purpose. In addition, industry, academe, and the professional societies need to join forces in developing rational approaches to facilitate gifts of laboratory equipment to colleges of engineering; one approach could be to promote legislation for this purpose where necessary.

The Two-Tiered System

Beginning in the 1950s the federal government initiated a comprehensive system of support for academic research and graduate education in the sciences. As the system grew, engineering research and graduate education began to be included. The objective was (and is) to develop knowledge and improve research techniques across a broad spectrum of disciplines, as well as to ensure a flow of graduate-level personnel to meet the nation's research needs. However, an unintended effect of this focused funding has been the creation of a two-tiered system of engineering colleges.

Rapid growth in funding took place during the 1950s and 1960s, followed by another upswing in the late 1970s that slowed to a modest increase in the 1980s. By 1981, federal government support for academic R&D was about $2 billion annually.

The impact of this comprehensive program of federal funding has been substantial. Three decades of rising annual funding fostered a group of research universities or institutions—the first-tier schools—whose graduate and research programs became heavily dependent on contract research. This system of government grants and contracts has greatly benefited many engineering colleges, but its focus has been almost exclusively at the graduate level. As a result, it has been the driving force in graduate engineering education. It has produced an array of sophisticated laboratories, so that some 15 to 20 schools now have one or more unique and cutting-edge laboratory facilities for research.

The rise of the government-funded research university also affected industrial support for engineering education. Many in industry believed that, because of large, continuing government funding, the universities were no longer interested in working with industry. Consequently, the industrial contribution to university R&D decreased slightly for a period after 1960. It later rose again; but considering the greatly increased government contribution, industry's share (on a percentage basis) was cut nearly in half between 1960 and 1981.

Recently, some major corporations have made sizable grants to a relatively small number of institutions. However, most of these initiatives have focused on the graduate research level and the same group of institutions that have been the primary recipients of government funding. Industrial support for academic R&D expenditures now amounts to about 4 percent of the total (although it is around 10 percent for engineering research) (National Science Board, 1982). Thus the federal government plays the dominant role in funding academic R&D.

The major recipients of government funds for graduate education and research enjoy a distinct advantage that influences both graduate and undergraduate engineering education at those institutions.

- Their recruitment of faculty is enhanced because the young assistant professor can continue working in a research environment similar to that experienced in graduate school. Their policies thereby sustain and perpetuate the academic value system.
- Teaching loads at research universities are relatively low, and a faculty member has a cadre of research assistants.
- The research infrastructure includes laboratory facilities, access to modern machine shops, and extensive library holdings, along with—most recently—extensive computer equipment.
- Typically, the benefits also include strong secretarial and technical support as well as ample travel funds.

Taken as a whole these benefits give a powerful impetus to academic research in graduate engineering education.

At the undergraduate level, no set of national policies or programs recognizes the important role of engineering education in contributing to the imperatives of a technology-based world economy. Because government and industry focus on research and graduate education, colleges that have as their primary focus undergraduate education in engineering have not enjoyed the advantages just described. They occupy a second tier within the engineering educational system.

Because approximately half of the B.S. engineering degrees are granted by colleges of the second tier, government, industry, and academe will continue to depend upon graduates of these primarily undergraduate colleges for at least half their engineering work force. Yet, because both government and industry focus their funding on graduate study and research, these colleges are forced to depend on other, appreciably smaller sources of funding.

In order to provide a measure of balance in this two-tiered system, the needs of primarily undergraduate institutions require recognition. Funding for modern laboratory equipment is an urgent need (see the section, "Equipment Obsolescence"). Colleges are experiencing a wave of computerization at the undergraduate level but most lack the resources to respond in a timely and comprehensive manner.

In addition, faculty who carry heavy undergraduate loads need support and access to creative programs of faculty development. Release time is especially valuable because it enables the individual to stay current in a professional field and develop new teaching techniques at the undergraduate level. Although the number of advanced academic research laboratories is limited, faculty members in primarily undergraduate programs nevertheless need access to major research centers in industry, government, and other universities in order to remain vital. Thus programs and policies are needed to enable these faculty members to take advantage of such facilities.

The separation in the two-tier system will widen unless both government and industry introduce imaginative programs accompanied by more than token support. Ways must be found to provide for more equitable distribution of the many benefits that accrue to first-tier schools. Such efforts need not entail much higher costs. For example, schools applying for government funding of major research facilities should be required to include a plan for involving outside faculty in research at the facility. Without strong public policy in support of a balanced system, undergraduate education will not be able to maintain the pace required to meet national economic and strategic objectives. (See the report of the Panel on Undergraduate Education for further discussion of this problem.)

Student Demographics

Given the traditional view of the engineering profession as a bastion of white males, the change in composition of the engineering student population in recent years has been dramatic (see Figure 8). The most noticeable change has been in the enrollment of women students, which has risen steadily in recent years from about 1 percent in 1970 to 15 percent (1983–1984) of the roughly 400,000 fun-time undergraduate engineering students nationwide (Engineering Manpower Commission, 1984a). However, the increase in percentage of women students may now be leveling off; it did not change substantially between 1983 and 1984.

The influx of women has been a significant factor in elevating engineering enrollments (and graduates) to their current high level. In 1970, for example, only 358 women graduated with bachelor's degrees in engineering. Ten years later there were 5,631 women graduates; the number rose to 6,357 in 1981, 8,140 in 1982, and 9,566 in 1983 (EMC, 1984b.). Yet considering that women constitute about 50 percent of the general population, they are still greatly underrepresented in engineering. As a result, many find engineering school to be a stressful environment in which they may experience a sense of isolation and a lack of acceptance on the part of faculty and male students (Report of the Panel on Undergraduate Education).

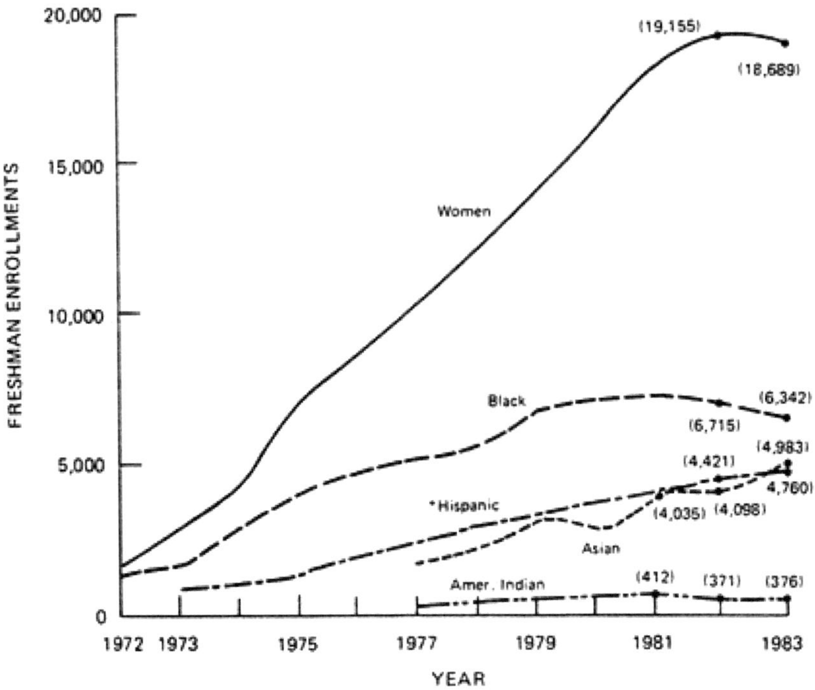

Figure 8
Freshman enrollments: Women and minorities.
Source: Engineering Manpower Commission, AAES.

Minority participation offers a similar picture in some respects. In the early 1970s, universities and colleges launched serious efforts to bring minorities into engineering; the efforts included scholarships and other types of financial aid, special academic programs, early recruiting, and the establishment of on-campus social support systems.

These efforts were successful to a certain extent, as Figure 8 illustrates. However, although the recruitment efforts continue, their effectiveness appears to have diminished. Except in the case of Asians, minority enrollments (i.e., of black, Hispanic, and American Indian students) had leveled off or begun to decline by 1982 (American Society for Engineering Education, 1983).

Much of the concern for minority enrollments has focused on the black community, which represents some 12 percent of the general population (Census, 1984). In 1982–1983, blacks accounted for only 4.4 percent of engineering students (EMC, 1983a). A variety of reasons for this limited participation have been given. One factor that is often mentioned is the limited exposure of predominantly inner-city black high school students to the idea of engineering as a profession, and to black engineer role models (see Report of the Panel on Engineering Employment Characteristics). Poor preparation in mathematics and science, limited funds, and a lack of self-confidence are also barriers to enrollment in engineering in many cases. Attrition among black students is higher than for any other demographic group, partly because of inadequate educational preparation and partly because of the social and economic factors just described (Report of the Panel on Undergraduate Education).

Enrollment of Hispanics and American Indians in engineering has also remained low in comparison to their numbers in the overall population, perhaps for much the same reasons.

As was just mentioned, the one exception to this low participation rate among minorities has been among Asian Americans. This group is represented in engineering schools at a disproportionately high level. While they make up only 1 percent of the general population (Census, 1984), they account for 3.9 percent of engineering students (1982–1983 data; EMC, 1983a). The group as a whole performs extremely well in engineering studies and tends to continue into graduate education (4.6 percent at the master's level and 4.3 percent in doctoral programs in 1982) at a higher rate than do other demographic groups (Report of the Panel on Undergraduate Education). A major factor in this performance is thought to be strong parental support for education in general and for scientific, mathematical, and technical pursuits in particular (Report of the Panel on Undergraduate Education).

The committee believes that the participation of women and minorities in engineering should be matters of continuing concern to the engineering community. The question of target levels of participation sometimes arises. Given the range of factors—some cultural, some social, some economic—that are outside the control of educators, it is probably fruitless to set such goals. However, some of the remaining obstacles can be identified and attempts made to reduce them. In this sense there is still much to be done.

For example, the quality of precollege preparation in science and mathematics has an important bearing on the participation of both women and minorities in engineering. For women, early exposure to physics appears to be particularly critical (Report of the Panel on Undergraduate Education). Poor preparation in these areas limits the appeal of engineering to these groups and increases attrition among those who do study engineering (especially among minority students). Educators should develop strategies to increase the size of the initial science/ mathematics pool of minorities and women.

Efforts must be made to reduce attrition of minorities all along the educational pipeline. For example, precollege programs such as those operating in a few major cities must be expanded and funded to prepare and motivate minority students to pursue college study and careers in engineering. In addition, mechanisms should be sought for providing needed social and academic support to both women and minorities in engineering education. Efforts such as these, vigorously pursued, can help to remove some of the invisible barriers that prevent the nation from gaining full access to the potential engineering talent embodied in large segments of the population.

By 1992, major demographic changes are very likely to cause a substantial drop in the number of qualified students entering engineering colleges in 38 states. Half of all B.S. graduates now come from 45 schools that have 400 or more graduates each year. Fourteen of those schools are in states (New York, Pennsylvania, and Massachusetts) where the high school population will decline about 40 percent by 1992. Twenty-seven of the 45 schools are concentrated in the 13 frost-belt states, which will all experience an appreciable decline (roughly 22 percent) in high school population (Report of the Panel on Undergraduate Education).

Some have suggested that the present high engineering enrollments (at 6 percent of all college students) represent a bubble, and that as the number of 18-year-olds declines, so will the number of engineering students. Data from 1983–1984 already show a decline of 6,000 in freshman engineering enrollments (EMC, 1984a). Certainly there will be regional effects of the differences in distribution of 18-year-olds. That is, schools will expand or shrink in size, some new ones will emerge, and others may close their doors, depending on changes in the size of the regional student pool.

To avoid being caught by surprise, engineering schools should examine the impact of prospective demographic changes in their area and should anticipate steps they will need to take to increase the flow of qualified students from their regional pool. Increasing the participation of women and minorities is one way to bolster enrollments. Other approaches will be specific to the circumstances of the individual institution.

Variability of Demand

Although natural market forces ensure a reasonably close balance between supply and demand for engineering graduates in different fields over time, occasional shortages and surpluses do develop. Currently, for example, because of rapid developments in microelectronics and the growth of information products, there are shortages in the computer and electronics fields. At the same time, because of the recent recession and shifts in our industrial base, there is less demand for civil and chemical engineers.

It is a mistake to overemphasize these current patterns of supply and demand. They are dynamic and change rapidly. However, their effect on the educational system is important. The perception of shortages and surpluses of engineers in certain fields (and the accompanying sense of excitement or disdain among students) has a dramatic impact on patterns of demand for particular courses of engineering study. Enrollments in electronic and computer engineering, for example, are saturated at most schools. The fact that the student response is usually out of proportion to the actual stimulus, combined with the fact that the response lags the stimulus by as much as four years, has the effect of wasting educational resources and engineering talent. Institutions cannot adapt to external conditions as rapidly as they develop; thus institutional stresses of this sort appear to have become a permanent feature of the contemporary educational environment.

It should be noted that many engineering educators do not consider the current overenrollments in electrical and computer engineering to be transitory. They believe that a structural change in engineering education is occurring—based on technological revolutions in these fields—that will keep demand high in the two disciplines indefinitely. Accordingly, some schools have considered instituting policy changes that would restrict entry into these fields of study on the basis of performance at different points along the educational path. Some have decided to do so (for example, the Georgia Institute of Technology, the University of California at Berkeley and at Davis, and the Massachusetts Institute of Technology). It remains to be seen whether such approaches will be workable and successful.

If further expansion in these and other high-demand disciplines is required, one approach would be to utilize dual-degree programs, also known as three-two programs. These involve a three-year generalist program (liberal arts, social sciences, mathematics and sciences) followed by two years of intensive engineering study, culminating in a B.A. /B.S. degree. Dual-degree relationships between liberal arts and engineering colleges have existed for at least two decades. They have enabled a few students, some from minority groups, to earn B.S. degrees in engineering. The capacity of the engineering educational system could be increased by creation of an explicit network of dual-degree programs, but such an approach would require a concomitant expansion in the two upper-class years as well. Dual-degree programs could be particularly effective at increasing the numbers of women and minority graduates. These dual-degree programs, in addition to the now-existing standard five-year track to the master's degree, would offer students a richer choice of options for their engineering education.

Continuing overenrollment in some disciplines exacerbates the faculty shortage problem. One way to achieve a degree of flexibility in dealing with the shortage on a short-term basis is to utilize professional personnel who are not in tenure tracks. Such individuals would include government, military and corporate retirees, with or without a Ph.D., who are not seeking tenure and who would welcome a short-term contract for a second career.

SPECIAL TOPICS

Specialization vs. Breadth

One possible way to ensure flexibility of response to fluctuating supply and demand and rapid technological change is to offer a broad engineering curriculum with many core engineering courses shared by students in all disciplines. This is not a new approach; various schools have found it effective over many years. Courses in the individual degree programs can be added or removed or modified in accordance with changes in the discipline and in professional objectives. The scope and content of these fundamental core courses will change with time as technology advances.

However, today there are forces militating strongly against this approach. The desire to provide specialized education at the undergraduate level has led to increasing fragmentation of the undergraduate curricula across engineering specialties. On the one hand is the requirement to prepare students with four-year degrees to be professional engineers; on the other hand is the need to provide a base for lifelong learning in specialties that may not yet exist. This increased specialization of engineering curricula, coupled with a decreased interest on the part of students in degrees in basic sciences and mathematics, will lead to future difficulties in our ability to respond quickly to new technological challenges.

The committee concludes that the undergraduate curriculum should provide considerable breadth across the disciplines of engineering and within each discipline. A broad engineering education leaves engineers better prepared to communicate with each other, to avoid technological obsolescence, and to learn new skills as technology advances. Extensive, in-depth disciplinary specialization does not belong in the undergraduate curriculum, and should be postponed to the graduate level. Neither is it possible in a four-year curriculum to treat all currently important technologies in a given engineering discipline.

Providing breadth of nontechnical education in the arts, humanities, social studies, and management also offers many advantages. Among the most important of these is an improved facility for communication, both written and oral. In an era in which communicating information has become a major component of virtually all professional work, the possession of good communication skills is increasingly important for engineers. However, the committee recognizes that it is difficult to provide even the approximately 15 percent of the time presently allotted for nontechnical breadth in today's accredited four-year programs. It is a perennial problem, and one with little hope of solution within present-day curriculum structures. Nonstandard educational tracks can produce at least some engineers having stronger nontechnical educational backgrounds. For example, with proper course selection, students who come into engineering in the context of a dual-degree program are better able to achieve this additional breadth. However, such programs cannot provide a complete answer. This topic is discussed further under the heading "Curriculum Requirements" in chapter 6.

Cooperative Education

One traditional aspect of the university-industry interaction is cooperative education, in which students hold part-time (during the school year) or full-time (alternating work and study) jobs in industry. Although cooperative education began over 75 years ago in the College of Engineering at the University of Cincinnati, only 2.5 percent (approximately 220,000) of the nation's 9,000,000 college students participate in co-op programs with 30,000 employers. Of the 404,000 engineering students nationwide, approximately 37,400 (or some 9 percent) participate in co-op programs. This national percentage is somewhat misleading, however, because many colleges do not offer the program at all. Where it is offered, student participation varies considerably from school to school. At some colleges, it is quite popular. For example, some 28 percent of the engineering students at Georgia Tech are co-op; at Northeastern University in Boston, the figure is around 80 percent.

These programs have provided a motivational component and a means of partly self-financing a college education. In addition, they give the student experience in observing the practice of engineering, an aspect that has been given less emphasis in contemporary engineering curricula. Thus they have an important orientational value, helping to enrich and focus the classroom learning experience.

Despite their usefulness, however, these programs entail additional administrative costs to schools, and have suffered from fluctuations in the economy and inconsistent support by industry. In addition, some educators express concern that co-op students have too little opportunity to socialize with other students and to participate in campus activities and are thus shortchanged in some very important nontechnical aspects of the educational experience. In reality, there is no reduction in on-campus time for co-op students; attendance during the summer is substituted for one of the other school terms. There is perhaps some disadvantage in this nonstandard enrollment pattern, but there is also a trade-off to be found in the closer student-faculty interaction that is possible in the overall co-op program.

Despite these concerns, the committee believes that in an educational environment characterized by constraints of various kinds, co-op education has a more important role to play than ever before. For example, the amount of project experience acquired by engineering students during their education has declined as the student/faculty ratio has risen. Thus, teamwork skills have suffered. Likewise, the hands-on experience base has suffered because of the shortage of laboratory equipment and instrumentation.

These educational shortcomings mean that graduates are not immediately valuable or productive when they enter industry; they require six months to a year of orientational training. The committee finds that some form of work experience during the period of schooling—whether acquired through co-op education, summer jobs, or some other form—is important as a means of offsetting these shortcomings.

To increase their effectiveness and enhance their role, co-op education and other such "interning" programs need to be strengthened. A considerably stronger commitment from industry and education is required to eliminate the boom or bust cyclical nature of support that tends to characterize these programs. Accordingly, the committee strongly recommends that the National Academy of Engineering and the professional societies take the initiative in bringing together representatives of industry, academe, and government to develop better work-study programs. Means should be found to eliminate the cyclical nature of support for these programs and to make it feasible for a much larger fraction of the engineering student cohort to participate.

Continuing Education and Professional Development

Considering the explosive growth in scientific and engineering knowledge since mid-century alone, it is worth noting that the average duration of engineering study has not increased substantially in that time. Even including those who receive the Ph.D., in a 30-year career after high school only 4 to 8 years consists of formal college education. During the remaining 22 to 26 years, education is obtained through a generally haphazard process of on-the-job learning, company training programs, seminars, conferences, and professional reading. It is estimated that only about 5 percent of this continuing education consists of formal classes or training programs (Report of the Panel on Continuing Education). Yet continuing education in all its forms is effectively the only line of defense for engineers against technological obsolescence brought about by changing technology.

Continuing education has not always enjoyed great popularity among companies or their employees. Chief executive officers and engineers alike have not generally understood its value. However, by 1977 over half of all practicing engineers were participating in some type of training activity each year. The two reasons given most often for this involvement are to prepare for increased responsibility or promotion and to acquire the ability to perform one's present job more effectively (Report of the Panel on Continuing Education). Obtaining credit toward an advanced degree is not the primary reason.

The underlying reasons for this growing emphasis on continuing education and professional development include the rapidity of technological change in every field of engineering, the introduction of computers (with their widespread impact on every discipline), the increasingly interdisciplinary nature of engineering work, and increased world competition in engineering requiring greater engineering performance. None of these underlying causes will disappear in the future. If our goal as a nation is to maintain a strong engineering work force, continuing education will have to play a vital role. Engineers can be productive over a longer period (thus expanding the engineering work force) if they have access to effective continuing education.

To meet the demand for continuing educational opportunities, new instructional sources have sprung up. A major provider is industry itself, which offers short courses and ongoing training programs in subject areas of interest to the individual company. The American Society for Training and Development (ASTD) estimates that in 1983 industry spent about $30 billion for all training and education—

although only a fraction of this amount was directed at engineering/ technical employees. Government also provides extensive education and training to its employees, at an estimated cost of $10 billion per year (Report of the Panel on Engineering Employment Characteristics). Professional and technical societies offer a broad selection of continuing education courses as part of their membership services. In addition, private vendors offering seminars and short courses on a broad range of engineering topics are now proliferating. The greatest demand is for highly targeted short courses that focus on new and developing technologies.

By and large, universities as institutions have not participated extensively in this activity, and when they do, it is not given much emphasis. Individual faculty members have been very active in providing courses through professional societies, as consultants, and as entrepreneurs. But to universities, continuing education means course work not intended for credit toward a degree; and the primary emphasis of universities is on undergraduate and graduate education. The provision of noncredit instruction is usually viewed as a public service.

However, some schools are finding that involvement in continuing education can be rewarding. Not only is it a source of income (however marginally), but it also increases the university's contacts with industry and its overall visibility. Perhaps the greatest potential for future expansion of continuing education is in the use of new educational technologies. A dedicated satellite link, for example, offers the opportunity for an interactive network of courses available at an engineer's home or place of work.[3] Self-paced and computer-aided instruction using microprocessors could become an efficient means of acquiring training. Such approaches lend themselves to the personalized and customized quality of continuing educational needs.

Some engineers can maintain their competence without structured education and training beyond college. However, most engineers will need continuing education throughout their careers if they are to remain competitive in the job market. Likewise, companies need their engineers to maintain competence if they are themselves to remain competitive in their markets. Continuing education is a unique field of engineering education that requires clear objectives and an increased understanding of its value if those needs are to be met.

---

[3] The National Technical University, now beginning operation, is an example of such a network (Baldwin, 1984a; 1984b).

Educational Technology

Applications of modern technology to education are often cited as promising ways for faculty to deliver more and better education to more students in less time. Undoubtedly, technologies such as the computer and satellite transmission have great potential—much of it still untapped despite exhortations over the years to schools and the government to provide for their greater use. However, there are three unsolved problems with this approach: the large initial capital cost, the reduction of student/faculty interaction with its concomitant cost in educational quality, and the fact that considerable faculty time is required for development (Report of the Panel on Undergraduate Education). Personal contact with a capable and experienced professor is an irreplaceable part of the educational experience. It is from such contact that students acquire a personal style of attacking engineering problems. That mentoring function is one that cannot easily be provided using new educational technologies.

However, there are many types of courses and many uses for which educational technology, properly implemented, offers great potential. The committee encourages engineering schools to create programs for development of educational technology by faculty, using shared institutional, industry, and government funding, and to implement these tools as fully as possible within their academic programs.

AREAS OF RAPID CHANGE

Student Prepareduess

One area of striking change in recent years has been the academic quality and ability of entering freshman engineering students. In sharp contrast to the late 1960s and early 1970s, when student interest in engineering was at its lowest point in decades, demand for engineering as a major is now extremely high. The result is that competition for places has been strong for several years, and engineering students nationwide are among the most able in their age cohort. This fact is illustrated by data for 1982, when, for the first time, average combined SAT scores of entering engineering students surpassed those of all non-science/mathematics majors (National Science Board, 1983).

Professors and employers alike refer to the dramatically higher communication and social skills of engineering students and recent graduates as compared to past stereotypes of the engineer. This trend may relate to a long-term shift in student socioeconomic levels overall. In the view of engineering deans and professors on the committee, today's engineering student (i.e., since the mid-1970s) tends increasingly to come from a middle-class, professional family background rather than the noncollege background that characterized many young engineers in the period after World War II. The predominance of such young people in engineering schools is now very strong. On balance, they have a richer educational and cultural background and are more confident, more assertive than engineering students of years past.

Another aspect of student quality relates to graduate students. Given the financial and other attractions that a career in industry offers to high-quality B.S. grads, one might expect to see a downward trend in academic quality among graduate-school applicants. However, that does not appear to be the case, because GRE scores of graduate applicants have remained relatively stable (Report of the Panel on Graduate Education and Research).

With regard to doctoral students, some academic administrators have reported that the academic quality (based on undergraduate class ranking) has fallen in recent years. However, there is some evidence that the trend is now reversing. Such trends are hard to verify, because it is difficult to obtain data on the problem from institutions, and because GRE scores are available only in the aggregate—rather than on the basis of doctoral candidate vs. master's candidate (Report of the Panel on Graduate Education and Research).

Despite the high ability of current engineering students, the committee is concerned that the erosion in precollege mathematics and science education, widely reported in recent years, threatens the base of the qualified engineering manpower pool (National Commission on Excellence in Education, 1983; National Science Board, 1984). This relates to an overall concern for the declining quality of secondary education, including written and oral communication skills. The engineering community must join in efforts to improve this situation.

Engineering Technology Programs

The period between about 1950 and 1980 saw a transformation of what were formerly called technical institutes or vocational schools into schools and colleges offering associate and bachelor's degree programs in engineering technology (Report of the Panel on Technology Education). The distinction revolves around the extent of formal mathematical and scientific training accorded to students in the newer programs, and the degree of technical sophistication and specialization required of graduates. Engineering technology curricula are in many ways similar to those found in engineering programs; the primary difference lies in a greater emphasis on applied practice and procedures in the former and a greater emphasis on fundamentals and theory in the latter.

There are areas of overlap in the work of engineers and technologists. Again, the primary distinction is one of a fundamental and theoretical focus versus an operational focus; engineers are usually involved in research, development, advanced design, and integrated design and manufacture, while technologists' work emphasizes known applications in design, manufacture, test, inspection, and quality control. The availability of well-trained engineering technologists is providing industry, at least in some sectors, with a greater flexibility in staffing. The outlook is for a greater output of technologists with more and more specialized skills to meet specific industry needs. Because of their training and the relatively close contact between schools of technology and their industry sponsors and clients, technologists tend to be of immediate utility to companies, thus reducing the training overhead burden.

Because of growing industry demand for these personnel, the number of institutions offering technology degrees has proliferated nationwide, from about 68 in 1951 to the current total of 154 accredited institutions offering two-and four-year degree programs (Report of the Panel on Technology Education). Many are community colleges offering a two-year transfer program leading to a bachelor of engineering technology degree at a four-year college. A large number of universities and colleges offer the four-year program. The popularity (and usefulness) of these programs is indicated by the fact that, between 1971 and 1983, the number of bachelor of engineering technology degrees awarded increased 79 percent (to 9,200) (Engineering Manpower Commission, 1984c). Enrollments do show a wide variability from year to year, however (EMC, 1983b). There is also great variability among engineering technology programs in terms of entry requirements, standards of achievement, curricula content, semester-hour requirements, and overall quality. More standardization in these programs could be achieved through interinstitutional cooperation.

A degree of friction has developed between engineering faculties and engineering technology faculties in universities offering both programs. The difficulty arises from the blurring of distinctions between the programs and from the competition for funds, laboratory equipment, and in some cases jobs for their graduates. Ultimately, the demand in the marketplace will determine the amount of emphasis that engineering technology education should receive.

Computer Science

Computer science is a rapidly emerging discipline, crucial to engineering. There is currently a great deal of variability in where computer science falls in the academic scheme of things, with computer science programs occupying a wide range of departments across different universities. Sometimes it is a part of engineering, sometimes in the mathematics department, and sometimes independent.

Two professional groups, the Institute of Electrical and Electronics Engineers and the Association for Computing Machinery, have recently joined in creating a special commission to consider the issue of accreditation for computer science programs.[4] Success in this effort should help to define more clearly the place of computer science as a professional discipline within university curriculums.

What is clear in any event is that contemporary engineering work in nearly every field requires some theoretical understanding of computers and programming. It is widely accepted that the use of computers must eventually pervade all fields of engineering education.

University-Industry Interactions

Under the pressure of foreign competition in engineering-intensive industries, the federal government has recently begun to encourage closer interactions between industry and universities (National Science Foundation, 1982b). In addition to direct support of joint research, various other steps that the government is taking will further improve the climate for university-industry interaction. For example, the administration has approved the concept of a closer collaboration between federal research laboratories and their university and industry counterparts (Office of Science and Technology Policy, 1983). In addition, the movement toward establishing up to 25 engineering research centers at engineering schools is encouraging (National Academy of Engineering, 1984).

State programs have also come to be very important in this regard. There are currently a number of fine examples, with North Carolina's Research Triangle Park being perhaps the best known. Others, at the University of Arizona, at Rensselaer Polytechnic Institute in New York State, and elsewhere, are becoming increasingly active. Such programs generate enthusiastic support in state legislatures and in localities

[4] The Computer Science Accreditation Commission, or CSAC.

because of the prestige, revenues, and jobs associated with them. They are also beneficial to engineering education at the participating schools in that they attract and stimulate highly qualified faculty and students, as well as industry funds and support.

Industry increasingly realizes that it has a crucial stake in the continued health of the engineering educational process, and in the quality of the educational product. Collaboration takes many forms. In some cases it is in the form of financial support through research grants to faculty and fellowships to graduate students, or through gifts of needed laboratory equipment. A growing trend is for the establishment of joint research endeavors between a university and a nearby company, either in the university research center or on-site in the company's laboratories (National Science Board, 1982). The federal R&D tax credit has been invaluable in helping to stimulate all these forms of industry support of research in engineering schools.

The use of adjunct faculty from industry to augment engineering faculties is a traditional concept, although its value is generally limited to instruction alone, and does not extend to full participation in other campus responsibilities. Similar, but with its own difficulties, is the concept of shared professorships, in which a faculty member and a practicing research engineer exchange places for an academic period. Faculty consulting to industry is also valuable in that it enhances university-industry contacts. Along with shared professorships, consulting offers the benefit of keeping faculty current with modern practice and the applications of research in the field. Consulting is therefore an important vehicle for feedback of ideas from industry into the classroom while providing industry with ideas based on academic research.

There are a number of actual and potential problems associated with university-industry interactions that that be satisfactorily addressed as those interactions become closer and more routine. One problem is based on the commercial nature of industrially sponsored research. Conflicts between the profit-making purposes of industry and the educational purposes of universities have to be resolved if productive collaboration is to occur.

As a general rule, the closer a university comes to the activity of product development, the less likely it is that the purposes of the university will be well served. Such activities are highly specialized, whereas the educational process should strive for generalizable knowledge. Secrecy constraints, often important to industry, are also in conflict with the generalizability of learning.

The ownership of intellectual property—usually meaning patents and copyrights—is another vexing problem for universities involved in industry research, especially in publicly supported universities. Similarly, consulting by faculty sometimes draws allegations of conflict of interest and inattention to the faculty member's teaching responsibilities. These issues are often heavily loaded with value judgments and political philosophies, yet they must be resolved if satisfactory university-industry relationships are to be developed. (See the report of the Panel on Graduate Education and Research for a more extensive discussion of these issues.)

This litany of concerns and issues regarding university-industry relations should not be too intimidating. In reality, there have been many instances of satisfactory relationships being worked out which maintain the integrity of the university's role while satisfying the requirements of the industrial organization.

FINDINGS, CONCLUSIONS, AND RECOMMENDATIONS

1. A broad engineering education leaves engineers better prepared to communicate with each other, to avoid technological obsolescence, and to learn new skills as technology advances.

   The undergraduate curriculum should provide considerable breadth across the engineering disciplines and within each discipline. Extensive, in-depth disciplinary specialization should be postponed to the graduate level.
2. Because few women chose to study engineering in the past, the profession lost access to substantial human resources. However, during the last decade the number of women studying and practicing engineering has increased dramatically, from 1 percent of engineering enrollment in 1970 to 15 percent in 1984.

   To achieve the full potential that this human resource offers, colleges of engineering and engineering technology, school systems, government, industry, and the engineering profession must continue to work to increase the number of qualified women who study for a career in engineering. The most important means are: greater effort (as recommended by other study groups) to increase the study of math and science by female secondary-school students and further action by colleges of engineering to increase female enrollment.
3. Blacks, Hispanics, and American Indians are greatly underrepresented in the pool of engineering school applicants (both graduate and undergraduate) and in the engineering workplace. This underrepresentation has social, economic, and educational origins. Despite recent

increases in minority enrollments, the potential representation of these populations remains unmet, and once admitted, their attrition is disproportionately larger than that of traditional engineering students.

Broader efforts by schools, companies, and engineering societies are required to bring more minorities into engineering. For example, precollege programs such as those operating in a few major cities and regions must be expanded and funded so as to better prepare and motivate minority students to pursue college study and careers in engineering. Retention programs similar to those now supported by many colleges and organizations must also be expanded.

4. Engineering co-op programs have traditionally filled a valuable role in engineering education. They provide a motivational component and a means of helping to self-finance a college education. In addition, they give the student experience in the practice of engineering, an aspect that has been given less emphasis in contemporary engineering curricula. Thus they have an important orientational value, helping to enrich and focus the classroom learning experience. Despite their usefulness, however, these and other such work-study programs (including summer employment) have traditionally suffered from fluctuations in the economy and generally inconsistent support by industry.

   To increase their effectiveness and enhance their role, co-op and other work-study programs need to be strengthened. A considerably stronger commitment from industry and education is required to eliminate the boom or bust cyclical nature of support that tends to characterize these programs. The committee strongly recommends that the National Academy of Engineering and the professional societies take the initiative in bringing together representatives of industry, academe, and government to develop better work-study programs. Means should be found to eliminate the problem of cyclical support and to make it feasible for a much larger fraction of the engineering student cohort to participate.

5. By 1992, major demographic changes will cause a substantial drop in the number of qualified students entering engineering colleges in 38 states. Half of all B.S. graduates now come from 45 schools that have 400 or more graduates each year. Fourteen of those schools are in states (New York, Pennsylvania, and Massachusetts) where the high school population will decline about 40 percent by 1992. Twenty-seven of the 45 schools are concentrated in the 13 frost-belt states, which will all experience an appreciable decline in high school population.

Engineering schools should examine the impact of prospective demographic changes in their area, in order to anticipate steps they will need to take to increase the flow of qualified students from their regional pool. Increasing the participation of qualified women and minorities is one means of bolstering enrollments. Other programs specific to the circumstances of the individual institution will also need to be devised.

6. Serious erosion of content and standards in virtually every area of study has occurred in secondary school systems over the last two decades. Critical shortages of science and mathematics teachers exist in almost every state. And half of the newly employed science and mathematics teachers are not qualified to teach these subjects. This erosion in mathematics and science, as well as in reading and writing, now threatens the base of the qualified engineering personnel pool.

   To improve the qualifications of students intending to study engineering, the schools—together with engineering education and professional societies—must actively encourage government and industry to join them in improving mathematics, science, technology, and communications content in secondary school curricula. The committee supports the recommendations put forth in recent studies by the National Commission on Excellence in Education and by the National Science Board's Commission on Pre-College Education in Mathematics, Science, and Technology.

7. The presence of a sufficient number of Ph.D. holders in the engineering work force will continue to be important, from the standpoint of both engineering research and teaching. Engineering Ph.D.s awarded are expected to increase to an estimated 4,000 per year by 1988. However, this increase will not be sufficient to meet requirements for additional faculty in the face of anticipated increases in industry demand and an insufficient proportion of U.S. residents in the Ph.D. student pool.

   A major increase in fellowship support and concomitant engineering college research support are needed to attract more of the very brightest U.S. citizens into graduate programs in engineering. To attract top students into graduate work, doctoral fellowships should carry stipends equal to at least half the starting salary of a new B.S. graduate.

8. The current and persistent shortage of faculty of sufficiently high quality is a serious problem for engineering education. Estimates of the

extent of the shortage range from 1,567 to about 6,700. (1,567 is the number of unfilled positions reported in a survey of engineering deans in 1983, and 6,700 is the number necessary to restore the student/ faculty ratio to that which existed in 1975–1976—often considered an optimal ratio.) The lack of sufficient high-quality faculty is the most important factor currently limiting attempts to increase the quality, scope, and number of engineering programs.

Increasing the supply of highly qualified U.S. residents holding the Ph.D. would help to alleviate the problem. (Restoration of the 1975–1976 student/faculty ratio, however, would require even further funding of graduate programs.) Universities, for their part, must make engineering faculty careers more attractive than at present in order to fill vacant faculty positions. Salaries need further improvement, adequate facilities are necessary, and current teaching overloads should be reduced.

9. Educational technology (computers, TV, satellite transmission, etc.) holds promise for improving the delivery of engineering education at all levels. However, the full implementation of educational technology has been inhibited by high costs and by the time required for faculty to integrate its use into the substance and process of the learning experience.

   Computers, and computer-aided instruction in particular, should be recognized as powerful educational systems tools. These tools should be applied as rapidly and as fully as practicable in all academic programs in such a way as to enhance the quality of engineering education. Engineering schools should be encouraged to create programs for development of educational technology by faculty, with shared institutional, industry, and government funding.

10. Engineers can be productive in engineering work over a longer period (thus increasing the size and effectiveness of the engineering work force) if they have access to effective continuing education. Needs of engineers for lifelong maintenance of competence through continuing education are met by a variety of means, including employers, professional/technical societies, academic institutions, private vendors, on-the-job learning, and the individual initiative of the engineer. However, the lack of company reimbursement and release time is a strong demotivator for pursuing continuing education.

    The various providers of continuing education should keep these educational sources available to the practicing engineer and should expand their offerings. Industry managers should recognize the value of continuing education in improving the effectiveness and adaptability of their engineering employees. Those companies that do not offer their engineering employees financial and worktime relief should strongly be encouraged to do so.

11. Industry's interest in engineering schools has traditionally focused on their product—the graduate. However, research in engineering in universities has become increasingly important to industry as well. In a climate of financial constraint and rising international competitiveness, industry has a vested interest in helping engineering schools to maintain high levels of educational and research quality.

    Closer ties should be fostered between university and industry. Creative and innovative ideas along the lines of the Semiconductor Research Corporation and the NSF's Engineering Research Centers are invaluable. In addition, current programs of industry-sponsored research, advisory councils, shared faculty, industry financial support for equipment and facilities, and joint industry-university provision of continuing education should all be encouraged. Continuation of the R&D tax credit is essential for maintaining all forms of industry support for research in engineering schools.

12. Laboratory equipment in engineering education has deteriorated over a long period of time. Plant and other facilities have also aged greatly. Governmental and industrial equipment support programs have been sporadic, so that a serious mismatch exists between the need for equipment and the level of support.

    A national program of government-industry-college matching grants is required to address this problem. Industry, academe, and the professional societies need to join forces in promoting legislation where necessary to facilitate gifts of laboratory equipment to colleges of engineering. In the special case of bricks and mortar, the federal government and industry should be prepared to match those funds raised by state governments or from philanthropic sources for this purpose.

13. There is great variability among engineering technology programs in terms of entry requirements, standards of achievement, curricula content, semester hours required, and overall quality. However, this diversity serves a useful purpose, given the diversity of industrial needs in different regions.

    Technical and technology institutions should cooperate in eliminating variability that has no relevance to market needs and is strictly arbitrary in nature.

14. Beginning in the 1950s the federal government developed a system of massive support for research and graduate education in science and engineering. This support led to a rapid growth of research institutions. At the undergraduate level, there has been no set of national policies or programs which recognizes the important role of undergraduate engineering education in contributing to the imperatives of a technology-based world economy. Because government and industry focus on research and graduate education, a two-tiered, or bifurcated, system of engineering colleges has been created. This two-tiered system has a strong influence on the character of engineering education. Government, industry, and academe will continue to depend on graduates of the primarily undergraduate-oriented colleges for at least half of their engineering work force. Yet, because both government and industry focus their funding on graduate study and research, these colleges are forced to depend on other, appreciatively smaller sources of funding.

    The federal government and industry should recognize and support innovative programs in undergraduate engineering education in the second-tier institutions. First, to ensure that the program quality of primarily undergraduate-oriented engineering colleges continues to meet the needs of a technology-based economy, these colleges must have access to new and additional sources of income. In addition, ways must be found to provide for more equitable distribution of the many benefits that accrue to first-tier schools. For example, faculty members and students at second-tier institutions will need to be involved with research facilities and programs of major centers of research.

15. Over many decades, the engineering educational system has adapted itself to relatively large fluctuations in enrollment. The elasticity of the system has been stretched to the point where it is now saturated in many disciplines. If further significant expansion is required, one way to achieve it would be to utilize dual-degree programs and transfer programs with community colleges. For at least two decades, a number of dual-degree relationships have existed between liberal arts and engineering colleges. These programs have enabled a modest number of students—some from minority groups—to earn B.S. degrees in engineering. The capacity of the engineering educational system could be expanded by creating an explicit network of dual-degree programs, but such a program would require a concomitant expansion of the two upper-class years of engineering education.

    The National Science Foundation should examine experience to date with dual-degree and other alternative engineering programs and should then take the initiative (if indicated) in establishing a pilot group of colleges and engineering schools to demonstrate effective structures for such programs. This pilot program could be funded by a combination of foundations, industry, and government agencies. Experience gained from the program could then be applied to a wider group of institutions. In addition, the experience gained would be relevant to the often-debated model of preprofessional followed by professional engineering education. It would also be highly relevant to the examination of options for restructuring the curriculum to meet competing educational demands. (see chapter 6, recommendation 7).

16. The shortage of faculty is likely to remain a serious problem. Although the issue of Ph.D. versus M.S. degree as a criterion has not been resolved, the Ph.D. has been a virtual requirement for tenure-track positions. To avoid this constraint, especially in times of faculty shortage, colleges of engineering can utilize professional personnel who are not in tenure-track positions.

    Engineering faculty members and administrators should identify and utilize as faculty individuals such as government, military and corporate retirees, with or without a Ph. D., who are not seeking tenure and who would welcome a short-term contract for a second career.

References

American Society for Engineering Education. 1984a. Policy statement endorsed by the Executive Committee of the Engineering Deans' Council .

American Society for Engineering Education. 1984b. Survey of engineering colleges.

Baldwin, L. V. 1984a. An electronic university, IEEE Spectrum (November), p. 99.

Baldwin, L. V. 1984b. Instructional television, IEEE Spectrum (November), p. 101.

Engineering Manpower Commission. 1983a. Engineering and technology enrollments, Fall 1982. Pt. I: Engineering. Washington, D.C.: AAES.

Engineering Manpower Commission. 1983b. Engineering and technology enrollments, Fall 1982. Pt. II: Technology. Washington, D.C.: AAES.

Engineering Manpower Commission. 1983c. Salaries of engineers in education.

Engineering Manpower Commission. 1983d. Salaries of engineers in industry.

Engineering Manpower Commission. 1984a. Engineering Manpower Bulletin, No. 73, July.

Engineering Manpower Commission. 1984b. Engineering and technology degrees—1983. Pt. II: By minorities. Washington, D.C.: AAES.

Engineering Manpower Commission. 1984c. Engineering and technology degrees—1983. Pt. I: By school. Washington, D.C.: AAES.

Lear, W. E. 1983. The state of engineering education. Journal of Metals (February), pp.48–51.

National Academy of Engineering. 1984. Guidelines for engineering research centers. Washington, D.C.: NAE.

National Association of State Universities and Land Grant Colleges. 1982. Report on the Quality of Engineering Education. Report of the Committee on the Quality of Engineering Education, of the Commission on Education for the Engineering Professions.

National Commission on Excellence in Education. 1983. A Nation at Risk. A report to the nation and the Secretary of Education.

National Science Board. 1982. University-Industry Research Relationships: Myths, Realities, and Potentials. Washington, D.C.: National Science Foundation.

National Science Board. 1983. Science Indicators: 1982. Washington, D.C.: National Science Foundation.

National Science Board. 1984. Educating Americans for the 21st century: A report to the American people and the National Science Board. Washington, D.C.: National Science Foundation.

National Society of Professional Engineers. 1982. Engineering Education Problems: A Guide to Action for NSPE State Societies. Washington, D.C. : NSPE.

Report of the Panel on Continuing Education, in preparation.

Report of the Panel on Engineering Employment Characteristics, in preparation.

Report of the Panel on Graduate Education and Research, in preparation.

5

Utilization of Engineering Resources

A major element of the integrative approach that the committee attempted to bring to the examination of contemporary engineering was to address the question of how members of the engineering community are employed in the workplace, and how engineering resources are utilized. The intent was not simply to include the study of utilization as an adjunct to the assessment of engineering education, but to view it as the other end of the pipeline, as part of the same system, and to attempt to highlight the interdependencies of the two.

As was mentioned at the beginning of the report, the subject of the utilization of engineers has not received nearly as much or as systematic a treatment in earlier studies as has education. Consequently, the Panel on Engineering Employment Characteristics, which examined this subject, was in many respects tilling new ground. The panel relied for its statistical data primarily on the same sources that were employed by the Panel on Infrastructure Diagramming and Modeling in its research (see chapter 3, "Data Bases").

Although the surveys conducted by these and other organizations supply a great deal of useful data, each agency collects information according to its specific needs and without reference to data from other sources or to a consistent set of definitions. This panel likewise found that the data bases, taken as a whole, exhibit numerous gaps and inconsistencies and are poorly suited to integrated analysis.

To augment the available information and to develop more current data on the utilization of engineers the panel also conducted an infor mal survey of employers of engineers. The survey was designed to yield an up-to-date picture of the quality of recent engineering graduates, the patterns of utilization of these personnel, and the impact of new tools on engineering productivity.

In accordance with the flow diagram of the engineering community developed by the Infrastructure group, the panel also sought to consider engineers, technologists, and technicians and to compare them in terms of employment and utilization characteristics. This section of the report examines

- Current characteristics of the engineering labor force
- Issues relating to the quality of the engineering work force from the standpoint of employers
- Current and future issues of supply and demand for technical personnel.

THE ENGINEERING WORK FORCE: CHARACTERISTICS AND TRENDS

According to Bureau of Labor Statistics data, between 1960 and 1982 the number of engineers in the United States nearly doubled, rising from 800,000 to about 1.6 million (Report of the Panel on Engineering Employment Characteristics). Figure 9 shows that the average rate of increase has also grown since 1976, a fact reflected in the high enrollments at engineering schools since the mid-1970s. Moreover, in the same 22-year period the number of engineers grew faster than the overall employed population. Engineers comprised nearly 1.4 percent of the United States work force in 1982, compared to 1.2 percent in 1960 (Report of the Panel on Engineering Employment Characteristics). [1] In recent years this growth has been especially strong in the manufacturing industries. Overall employment in these industries grew less than 3 percent during 1977–1980, while engineering employment climbed 20 percent (National Science Foundation, 1982a). Even in mature industries with declining employment, engineering employment remained relatively stable. In fact, some 75 percent of engineers work in industry and business (NSF, 1982a). These trends reflect both the spread of high technology throughout industry and the efforts of older industries to upgrade their productivity and competitiveness.

[1] However, because of a large increase in employment in non-engineering-intensive portions of the economy (i.e., the service sector), engineering employment as a percent of the work force has declined from a peak of 1.6 percent in 1970.

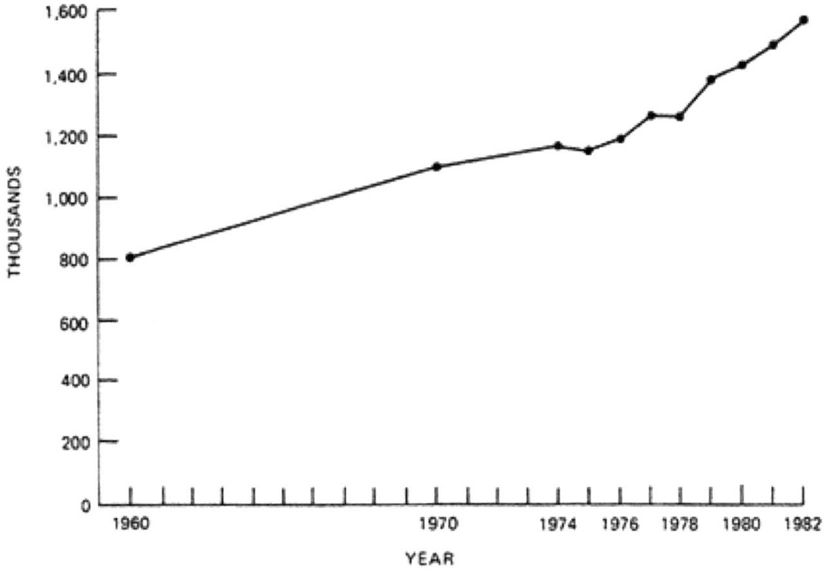

Figure 9
Employed engineering personnel: 1960–1982.

Concentration Ratios

One measure of the technology-intensiveness of an economic sector or industry is the proportion, or concentration ratio, of technically employed people in its total work force. (Figures in this section are again based on data from the Bureau of Labor Statistics.) Of the major economic sectors, for example, the federal government has the highest concentration ratio for engineers. The ratio rose from about 3.25 percent in 1960 to about 5 percent in 1978 (the latest year for which data are available). About 6 percent of all engineers are employed directly by the federal government (Report of the Panel on Engineering Employment Characteristics). When indirect employment is taken into account (i.e., prime contractors), the federal government employs some 30 percent of the engineering pool; second-tier indirect employment via subcontractors adds another 8 percent to the total. (Although these figures may seem surprisingly large, they are roughly equivalent to the portion of the overall GNP accounted for by the federal government.)

The concentration ratios for engineers in other sectors are considerably lower: durable goods, 4 percent in 1978, with the trend being downward; nondurable goods, slightly over 1 percent in 1978, with no change expected in the near term.

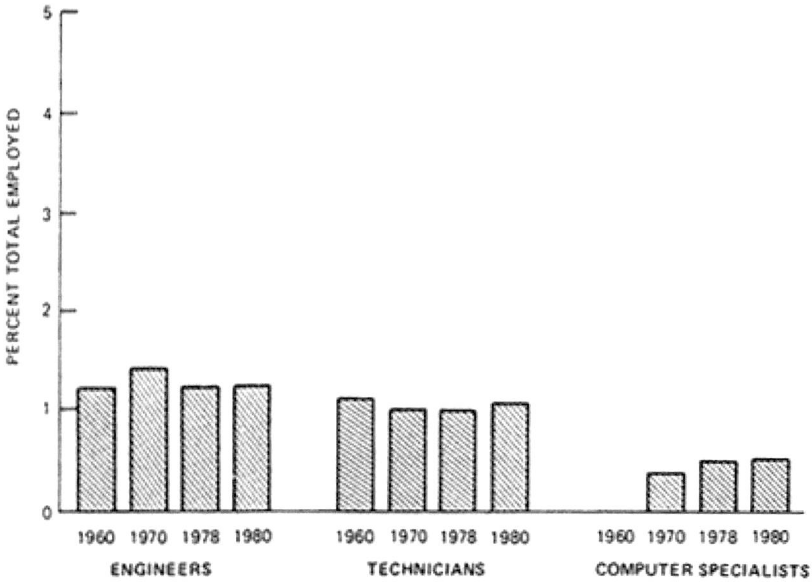

Figure 10
Engineerings, technicians, and computer specialists as a percent of total employed: All industries.
Note: 1980 figures are estimated.
Source: Bureau of Labor Statistics.

Ratios vary widely across industries. They are highest in manufacturing industries generally, as might be expected, although the highest ratio, 22.7 percent, is found in the engineering services industry. Next highest are the aerospace industry 13.85 percent), commercial R&D (12.1 percent), computers (9.2 percent), and electrical machinery (7.0 percent).

Concentration ratios for engineers, technicians,[2] and computer specialists in all industries are compared in Figure 10. It should be noted that engineers (as defined) outnumber technicians—and these figures include not only engineering technicians, but scientific technicians as well. In 1982 there were 1.1 million technicians of all types in the total work force, compared to the nearly 1.6 million engineers. Among eco

[2] The category of technicians does not include those technicians who are performing professional-level engineering work and who are thus defined as engineers.

nomic sectors, the number of technicians (and thus the concentration ratio) exceeds that of engineers only in nondurable goods (e.g., fertilizers and food products). Among industries, the technician ratio is higher only in chemicals, engineering services, and commercial R&D. Computer specialists are a fast-growing category, but they currently outnumber engineers and technicians only in electronic computing and computer programming.

It is difficult to find accurate employment data on engineering technologists per se because the field is relatively new and because technologists are often classified by their employers and by themselves as engineers. Another factor is the relatively low number of technology schools reporting on enrollments and graduates. However, if the total number of baccalaureate technology degrees awarded each year is around 9,200 (as it was reported to be in 1983), then the yearly output of technologists is about 13 percent of the yearly output of new B.S. engineers (72,500 in 1983) (Engineering Manpower Commission, 1984a). Therefore, since there are relatively few older technologists, the concentration ratios of these employees must be considerably lower than those of engineers, even in the manufacturing industries where they predominate (see Figure 10).

The finding that there are apparently far fewer technicians and technologists in the work force than there are engineers was initially troubling because as it seemed to imply an inefficient use of resources. However, the committee found that self-reporting of data distorts the picture considerably (that is, many technicians and most technologists define themselves as engineers). In addition, there are many engineers who do technician-level work. Thus, there is a built-in asymmetry in the data for these groups. The occupational structure is actually not as top-heavy as it would appear to be. However, periodic monitoring of the situation would be advisable as one means of ensuring that engineering resources continue to be utilized efficiently.

Predominant Work Activity

By far the largest number of engineers are employed in the durable goods sector, which accounted for 40 percent of all engineers in 1978 (Report of the Panel on Engineering Employment Characteristics). However, this percentage is decreasing steadily while the proportion of engineers in the service sector grows. The continuing predominance of manufacturing employment nevertheless is reflected in the fact that across all types of employers the most frequent activities of employed engineers (in 1982) were development, production/inspection, and management (see Table 3).

TABLE 3 Primary Activities of Employed Engineers, 1982

Activity	Women Engineers (percent)	All Engineers (percent)
Research	10.9	4.7
Development[a]	15.2	27.9
R&D Management	3.4	8.7
Other Management	16.6	19.3
Teaching	7.3	2.1
Production/Inspection	13.6	16.6
Other[b]	33.0	20.7

NOTE: These data are compiled by NSF's National Science Board from a variety of sources, including employer surveys and engineer (self-reporting) questionnaires. Thus they reflect a considerable degree of subjectivity and inconsistency in the definition of activities.

[a] This category includes design activity.

[b] Includes consulting, reporting, statistical work, computing, other, no report.

SOURCE: Unpublished tabulations, National Science Foundation. Based on 1982 Post-censal Survey of Scientists and Engineers, July 1984.

The predominant activities of engineers on-the-job differ from those of scientists in the same industries. Scientists are more likely to be involved in research, analysis, and teaching. Even of those engineers employed by educational institutions, only about half are actually engaged in teaching. The rest are involved in such activities as R&D, administration, and facilities engineering.

Technologists and technicians are commonly viewed as working in support of engineers, but in fact the association is frequently indirect. Often they perform tasks such as testing, inspection, and quality control in which engineering specifications are followed but engineers themselves are seldom involved. New technologies are also creating jobs that did not exist before that technologists or technicians carry out without direct supervision by engineers. Some examples are CAD/CAM operator/drafter, operation of numerically controlled machine tools, and robotics supervision (Office of Technology Assessment, 1984).

Specializations

In 1981 the largest engineering disciplines were electrical/electronic and mechanical engineering. Table 4 gives the numbers and percentages of practitioners in the six largest disciplines, out of approximately 1.5 million employed in that year.

Since 1960 the fastest-growing categories have been the electrical/ electronics and industrial engineering disciplines. Figure 11 depicts these relative growth rates, using data from the Bureau of Labor Statistics. (Note that the curves do not reflect absolute numbers of practitioners.) The rapid growth in the "other" category, as shown in the figure, reflects the recent emergence of engineering fields such as environmental engineering and biochemical engineering (Report of the Panel on Engineering Employment Characteristics).

TABLE 4 Distribution of Engineers Employed in Six Largest Disciplines, 1981

	Engineers Employed	
Discipline	Number	Percent
Electrical/Electronic	279,200	18.9
Mechanical	249,500	16.9
Civil	200,300	13.5
Industrial[a]	143,000	9.7
Chemical	79,400	5.4
Aero/Astro	50,200	3.4

NOTE: Totals do not add to 100 percent because of the large number of smaller disciplines.
[a] Based on 1980 data adjusted upward.
SOURCE: National Science Board, 1983.

The growth in electrical/electronics engineering has been widely observed and is, of course, the result of breakthroughs in the development and application of microelectronics and computers (see Figure 11). The steady growth in industrial engineering is a consequence of industry's efforts to improve productivity, product quality, and cost-competitiveness. Industrial engineering is a good example of a field in which many practitioners are technologists, upgraded technicians, or individuals with technical degrees in other fields—a fact which is reflected in its large size relative to B.S. engineering degree output.

Women in Engineering

Women continue to be underrepresented in engineering. This conclusion is based on the committee's finding that the percentage of women is markedly lower in engineering than in other science and technical fields. While some 20 percent of chemists and 29 percent of computer specialists, for example, are women, only 5.8 percent of engineers are women (Report of the Panel on Engineering Employment Characteristics). However, the percentage of women in engineering practice more than tripled between 1970 and 1983, and the disparity in female representation in engineering now shows signs of rapid improvement. As a case in point, some 15 percent of undergraduate engineering students are now women; freshman female enrollments are even higher—17 percent in 1983 —although there are indications that the latter trend is leveling off (Engineering Manpower Commission, 1983; 1984b).

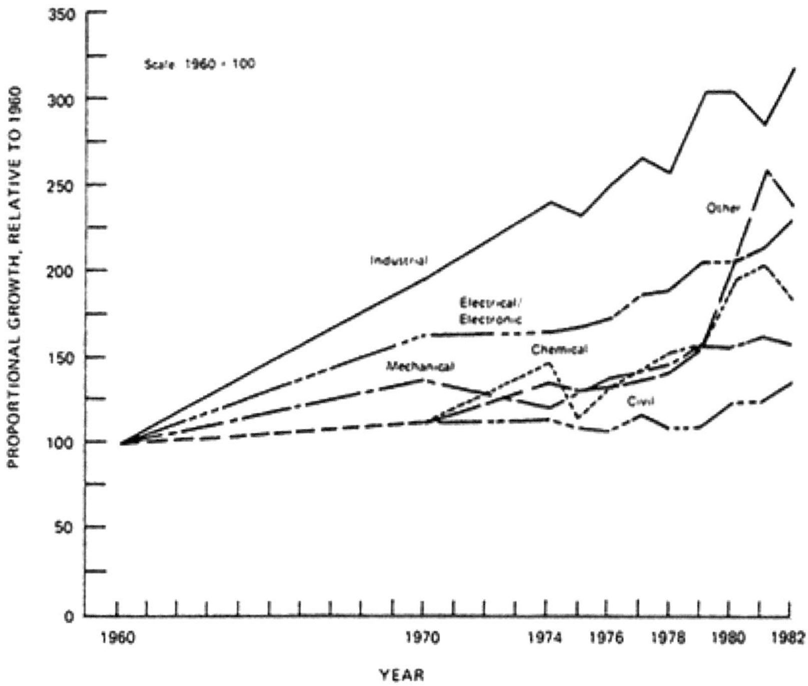

Figure 11
Relative growth rates of engineering disciplines, 1960–1982.

Of the major engineering disciplines in 1982, civil engineering had the largest proportion of women practitioners (12 percent); 11.7 percent were in electrical and electronics; 11.7 percent were in mechanical; and 11 percent were in chemical engineering. The percentage of women engineers engaged in research (10.9 percent) is more than twice that for men (4.7 percent), and the percent in teaching (7.3 percent) is more than three times that of men (2.1 percent).

While we have no reports of undue resistance in hiring or on-the-job discrimination by male coworkers or supervisors, it is obvious that some women will experience discomfort in an environment substantially populated by men. There is a relative scarcity of women in middle and upper management positions, but this could reflect the fact that women engineers are still too few and predominantly too young to be in competition for those positions. In addition, two recent reports point out that women engineers are paid 10 to 20 percent less than their male counterparts with the same experience—although neither report presents its findings as being conclusive (Institute of Electrical and Electronic Engineers, 1984; National Science Foundation, 1984). Other data indicate that women's entry-level salaries, at least, are substantially the same as those of men.

Anecdotal reports on the progress of women in engineering education suggest that female engineering professors are not obtaining tenure at the same rates as are their male counterparts (Report of the Panel on Graduate Education and Research). There is also a perception of discrimination against female faculty members in assignment of teaching responsibilities and in selection for research teams. Such a perception discourages women from entering graduate school and then academia—certainly an undesirable result in view of the current shortage of faculty. College administrators should make a candid assessment of the negative aspects of campus life for women faculty members and, if they are found to exist, should take firm steps to eliminate them.

Minorities

Minorities made up 4.6 percent of employed engineers in 1981. The largest minority grouping was Asians, which increased by 45 percent between 1976 and 1981, to 2.8 percent (or 41,800) of all employed engineers. The number of black engineers nearly doubled during that period, but still constitutes only 1.4 percent (or 20,600) of employed engineers. Hispanics were even less well represented, making up 0.3 percent (or some 5,000) of employed engineers in 1981. The number of American Indians employed as engineers was very small (National Science Board, 1983).

Some of the possible reasons for this disappointingly low participation by minorities were discussed in the previous section on the status of engineering education—particularly with regard to blacks. On-thejob, cultural factors play a large part in that minority engineers must still cope with a considerable degree of isolation in a work world in which they are ethnically almost alone. In many localities, minorities in certain professions (medicine, law, etc.) can serve their own ethnic communities. There is no such parallel professional engineering establishment serving the minority communities. That fact may steer many professional-minded minorities away from engineering.

Also, there are questions regarding the upward mobility of minori ties. However, as is the case with women, the relative newness and low numbers of minorities in engineering are certainly factors in their underrepresentation in management positions.

It may be that, as was seen in the case of women, the fuller participation of blacks and other minorities in engineering will be a process that is slow to develop but quick to accelerate when the necessary conditions are created. Consequently, the search for ways to encourage minorities to enter and remain in engineering must continue.

QUALITY OF THE ENGINEERING WORK FORCE

One of the most critical characteristics of a work force is its quality. But quality is invariably a matter of perception; its assessment depends on personal experience and personal criteria. Many observers in recent years have expressed their concern that the quality of the engineering work force in the United States is declining. These commentators point to problems in the nuclear power industry, recalls of automobiles, and the general decline of our smokestack industries as symptoms of poor engineering quality.

On the face of it, it seems unwarranted to blame engineering for these signs of widespread industrial malaise. Industrial decline has many interrelated causes. Certainly among the most prominent are shortsighted management, national priorities, economies in production made possible for competitors abroad by relatively cheap labor, and less stringent environmental regulations in many countries abroad. Nevertheless, just as sound engineering is essential to industrial success, inadequate engineering must eventually be reflected in industrial decline.

But it would seem to follow that the recent sustained improvement in economic indicators, the apparently successful retooling of the auto industry, and the continued strength and competitiveness of the U.S. electronics industry all owe something to high-quality engineering. To acquire some sense of the present and future quality of the engineering work force, the panel asked its survey respondents to characterize the most recent graduates in terms of quality.[3]

The majority of respondents noted an upward trend in the quality of graduates, with few respondents reporting declines in quality. A sub

[3] Survey questionnaires were mailed to 350 engineering-based firms across the country. A total of 107 responses were received. Findings based on the survey should be viewed in the light of this relatively small sample size.

stantial increase in the quality of computer hardware and computer software engineers was noted (Report of the Panel on Engineering Employment Characteristics). These findings, although they are subjective, may reflect the greater intrinsic ability of engineering students that was described in the section on education. That is, it is difficult to say whether the assessments of quality refer entirely to technical training and knowledge or whether they include an acknowledgment of the fact that these graduates are simply brighter and more well-rounded than may have been the case in the past. Certainly the current overcrowding of classrooms and obsolescence of teaching equipment must be limiting the educational quality that might otherwise be expected in these graduates.

Despite the satisfaction with the overall ability of recent graduates, most companies find that they lack the ability to step into a job and become immediately productive. Often, additional training of six months to a year or more is required to properly acclimate the new employee to the requirements of the job. Offering this finishing training is a particular problem for smaller companies because of its high cost.

The crux of the problem is that to make the transition from a high school graduate to a competent practicing engineer requires more than just the acquisition of technical skills and knowledge. It also requires a complex set of group-interaction, management, and work-orientational skills. Other very important skills are those needed for communicating effectively, both orally and in writing. These skills are not sufficiently emphasized in the educational background of most recent engineering graduates.

New technologies can improve both the productivity of engineers and the quality of their work. For example, computer-aided design (CAD) unquestionably increases an engineer's productivity in terms of hourly output (by as much as 50 percent, according to the limited survey in Report of the Panel on Engineering Employment Characteristics). However, it is misleading to assign a number to the productivity increase, because CAD also changes the nature of the work. It may permit the engineer to design a part with greater precision, for example, or to look at 10 design options instead of 2 within the same period of time. Also, designing with CAD facilitates the handling of routine tasks and permits engineers to more fully exercise their engineering skills, concentrating on more complex design questions. The resulting gain in efficiency is difficult to quantify, but is nonetheless real.

Although CAD relates mainly to engineering work in the manufacturing industries, the use of computers and computerized tools in general is having a comparable impact in virtually every field of engineering.

ISSUES OF SUPPLY AND DEMAND

Several panels looked at the subject of supply and demand for engineers from different points of view. The Panel on Engineering Interactions With Society took a historical overview to try to identify some of the dynamics and mechanical features inherent in the societal expression of demand and the engineering profession's response to it. The Panel on Infrastructure Diagramming and Modeling examined the issue from a systems standpoint, and attempted to itemize the flowpaths that characterize the response to demand. The Panel on Engineering Employment Characteristics took a general look at the subject. And, finally, the panels on Undergraduate and Graduate Education both examined the elements of the supply response.

All of these efforts led, from separate directions, to the conclusion that it is impossible to design systems for predicting or managing supply and demand for engineers in any meaningful way. The limiting factor is our ability to forecast developments such as levels of economic activity and capital expenditure, national priorities, and societal and world events in general. The impact of certain events in isolation can be predicted in rough terms, and the interaction between individual elements of the supply-demand system can sometimes be forecast. But even the best available model using the most rigorous description possible does not provide a scientific level of predictive capability.

However, it can certainly be instructive to examine piecemeal some of the factors that bear on the issue of supply and demand. One of these is the occurrence of shortages and surpluses of engineers in different fields.

Shortages and Surpluses

The past few years have seen frequent reports of shortages of engineers, notwithstanding the dampening effects of the recession of 1981–1982. Actual shortages, however, appear to have been limited to certain specialties such as electrical, electronics, and computer engineering, in which industrial growth continued to be strong (Report of the Panel on Engineering Employment Characteristics).

Some observers are concerned that shortages of engineers will persist beyond the near term, but one authoritative source (Bureau of Labor Statistics) expects problems only in those specialties involved in fast changing technologies. On the whole, BLS foresees an overall balance of supply and demand for engineers throughout the 1980s (Slaughter, 1981). However, the BLS predictions are based on a balance achieved through the adjustment of supply, including continued high levels of participation by women. Some academics are concerned that problems in the educational system (i.e., faculty shortages and outdated, inadequate facilities) could, unless properly and promptly addressed, affect their ability to provide adequate numbers of high-quality graduates. Furthermore, it is misleading to refer to an overall balance between supply and demand because the difference between stringent shortage and painful surplus in any discipline is about 5 percent in either direction.

Spot surpluses have also existed in recent years, although these have not received as much attention. Chemical engineering has felt the impact of surpluses because of economic downturn, decreased demand for petroleum-based products, and reduction in support of alternate energy programs. Civil engineers have likewise been in oversupply as a result of the impact of recession on the construction industry and of a lessened demand for environmentally related work.

It is important to emphasize that neither of these conditions (surplus or shortage) is static; they vary across time and in each discipline somewhat independently. A major initiative to rebuild the nation's aging network of highways and bridges could rapidly increase the demand for civil engineers, for example. Consolidation among the producers of electronics goods or successful entry of low-cost foreign producers could reduce demand for electrical, electronics, and computer engineers in this country. Change in the patterns of demand will certainly be seen, and it is likely to occur more rapidly than in the past.

Salaries

One indicator of demand for engineers is their salaries. The most recent earnings surveys show that engineers in industry remain among the best paid of all non-self-employed professionals. Figure 12 shows that industry-employed engineers as a group earn more than chemists and accountants and that since 1963 the percentage differential has remained essentially the same (Bureau of Labor Statistics, 1983).

The comparison for entry-level engineers is similar. They earn more than their counterparts in other fields, and after about 1977 the differential began to increase noticeably (Figure 13). By March 1984 the average entry-level B.S. engineer was earning $25,750, considerably more than entry-level employees in the other fields (College Placement Council, 1984).

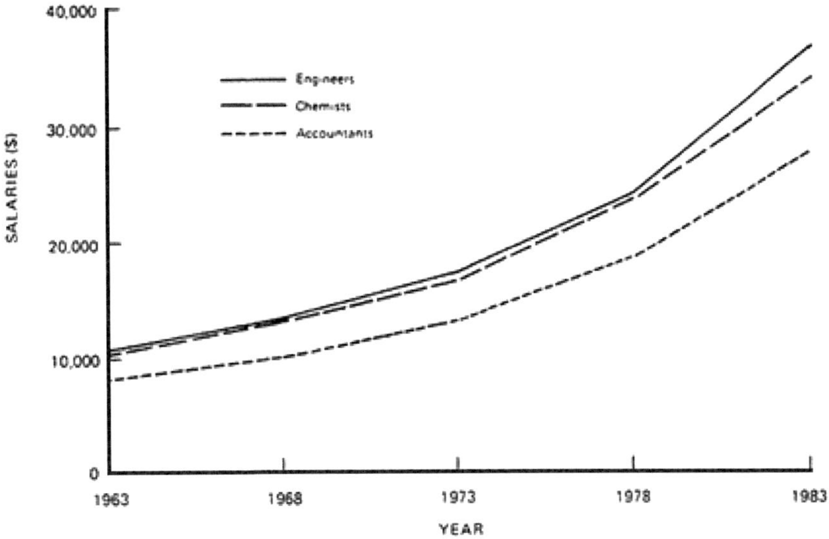

Figure 12
Median salaries for selected occupations in private industry (1963–1983).

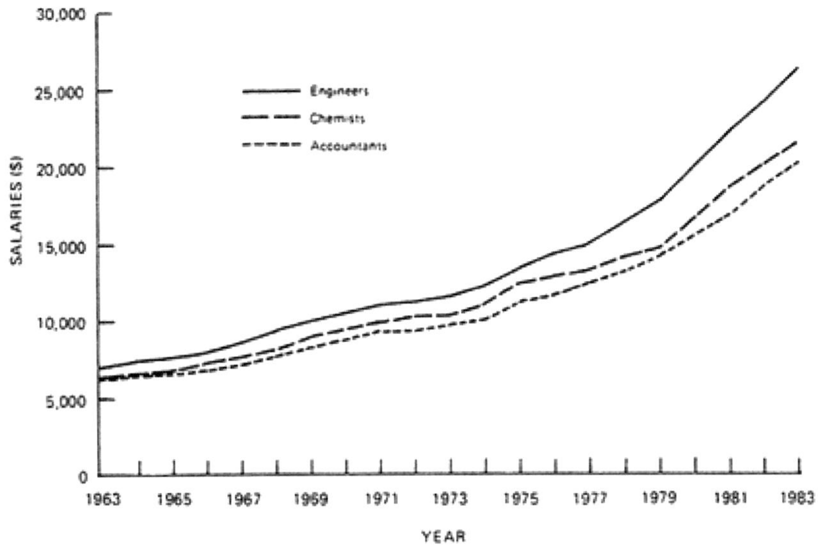

Figure 13
Entry-level median salaries in private industry for selected occupations (1963–1983).

The increase in actual salary differential suggests that employers considered engineers to be in short supply after 1977. Recent reports suggest that entry-level salaries in 1984 have begun to level off (Report of the Panel on Graduate Education and Research); if true, it would corroborate the earlier assertion that spot shortages are being filled. However, within narrow bands salaries may not be a particularly accurate index of demand; entry-level salaries paid to chemical and nuclear engineers, for example—two specialties in which demand has been low in recent years—are among the highest in any category (College Placement Council, 1984).

It should be noted that demand for degreed engineering technologists appears to be driving their starting salaries up to a level comparable to that of engineers. By early 1984 the average starting salary offer to a bachelor of engineering technology was $24,730, just $1,000 lower than the average offer to a B.S. engineer (College Placement Council, 1984).

Salary data also shed light on the relative reluctance of engineering students to pursue the Ph.D. Rough calculations by the committee suggest that a Ph.D. engineer does not surpass the total accumulated earnings of a B.S. engineer until about 21 years after each has received the B.S. (see Figure 14).

The salaries paid by industry for Ph.D.s are said to be a major lure for academic scientists and engineers alike. As was discussed in the section on faculty shortages, the disparity between engineering faculty and industry income is considerable, particularly for younger faculty members.

As is the case in universities, the federal government pays engineers at most experience levels and in most disciplines less than they can earn in industry. Federal salaries are limited by civil service regulations, and the salary differences—particularly at the higher levels—can be dramatic. Lower-level engineer salaries are also considerably below those in industry and are a major reason for the difficulty that government has in hiring engineers out of college. However, as in universities, government employment also has some offsetting benefits. Employment security, early responsibility, and the civil service retirement program have traditionally led the list (although the latter situation is now changing).

In view of the strong direct dependency on engineering talent for many of its most important activities, the federal government should review its compensation policies to ensure that it can competitively recruit and maintain a high-quality engineering work force.

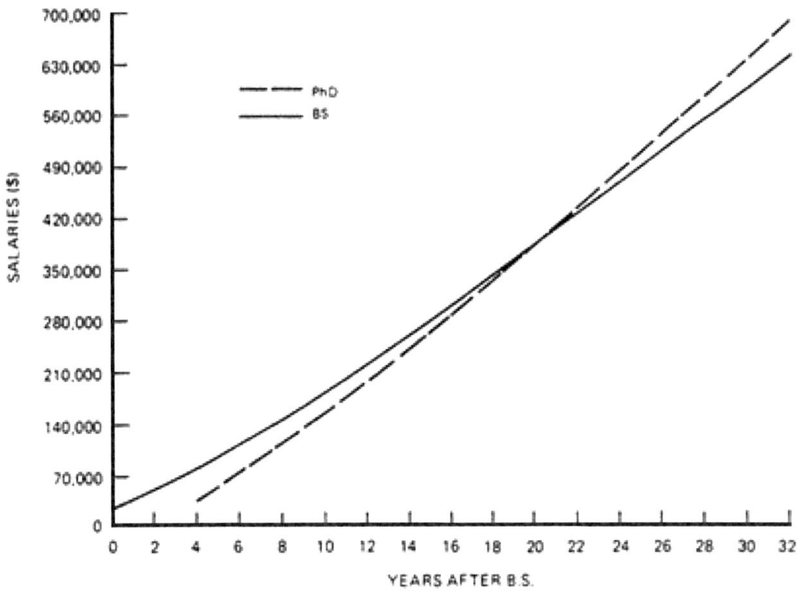

Figure 14
Cumulative B.S./Ph.D. Salaries.

Unemployment Rates

Another indicator of demand for engineers is unemployment rates. The rate for engineers traditionally has been markedly lower than for the labor force as a whole. Between 1963 and 1982, unemployment among engineers exceeded 2 percent in only four years; the rate peaked at 2.9 percent in 1971 (when aerospace cutbacks were most deeply felt) but hovered around 1 percent throughout most of the period. The rate in 1980 was 1 percent, compared to 7.1 percent for the labor force as a whole; in the same year it was 1.8 percent for physical scientists and 1.6 percent for social scientists (Report of the Panel on Engineering Employment Characteristics).

Although unemployment rates for engineers (as well as other professionals) may be understated somewhat because they are self-reported, it is nevertheless clear that engineers as a whole are seldom out of work.

Mobility

Another explanation for the low unemployment rates among engineers may be their mobility, both across fields and into and out of engineering. Data on the mobility of experienced engineers show a net flow of 18.5 percent out of the field during the period 1972–1978, corresponding to the highest unemployment years (Report of the Panel on Engineering Employment Characteristics). The data depict a net flow into management, a net flow out of production and R&D, and a small net flow out of teaching during those years. Later data show a small net flow out of teaching during 1980–1981 and a small net flow into teaching the following year (Geils, 1983). Engineers frequently move internally within a company to gain broader experience. The most common move is from one assignment to another at the same location. Engineers may also move (or be moved) geographically to take a new position or obtain a range of experience at different facilities of the same company.

Aging and Retirement

Another supply-side factor in the supply-demand equation is aging and retirement of engineers. The data on age distribution presage no age-related shortage of engineers overall; the greatest number of engineers today are in the 30–34 age bracket, while the average age is 42–44 (Figure 15). Data on specific disciplines do suggest that the nation faces a potential age-related shortage of experienced mechanical engineers when those now in the 45–55 age bracket begin to retire, unless demand drops proportionately (Report of the Panel on Engineering Employment Characteristics).

One ameliorating factor in the retirement equation is that engineers who retire do not necessarily stop working. Retired engineers commonly work as consultants, part-time employees, teachers, and so on.

THE IMPORTANCE OF ADAPTABILITY

Adaptability of Engineers

The research of all the panels demonstrated that adaptability to changing demand has been, and is, one of the most valuable characteristics of the engineering community—both individually and on the whole. This large, highly specialized work force has shown a remarkable capacity to adapt to fluctuating national needs while retaining the vitality needed to meet those challenges.

This capacity for adaptation is often in evidence when new technologies are introduced. A dramatic example was the substitution of tran sistors for vacuum tube technology in the mid-1950s, followed in the next decade by the substitution of the integrated circuit for transistors.

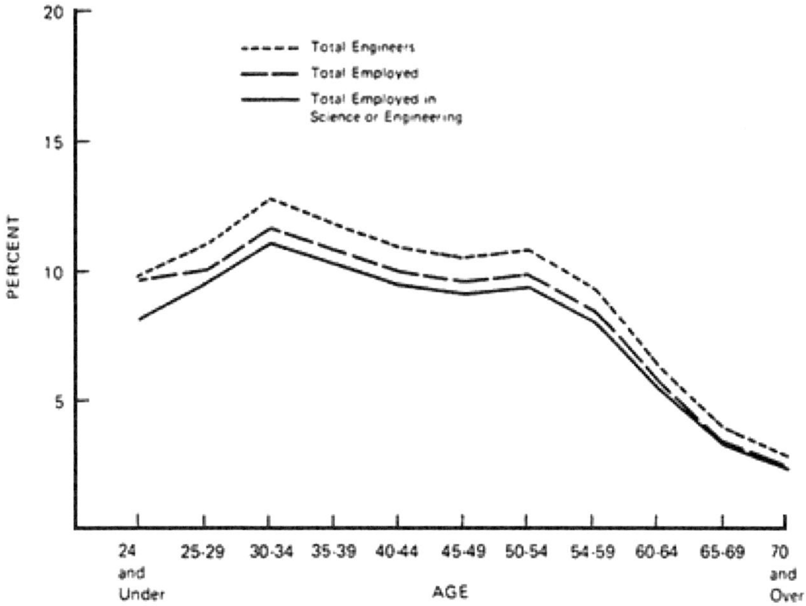

Figure 15
Age distribution of engineers.

Contrary to what might have been expected, the impact on engineers of those two events was relatively minor. In each case, the fact that there were virtually no engineers trained in the new technologies—and that the changes came so quickly—meant that practitioners of the obsolete technology were the best positioned and best prepared to apply the new technology. They adapted (Report of the Panel on Engineering Interactions with Society).

A different form of resiliency is seen when cross-disciplinary movement is required. For example, when the manned space program geared up in the late 1950s, there were virtually no qualified aerospace engineers. Instead, aeronautical, mechanical, and electronics engineers, mathematicians, and scientists of all types were able to adapt their knowledge to the requirements of the spaceflight regime. When the Apollo program ended rather abruptly in the early 1970s, those several thousand engineers were eventually reabsorbed by industry—although the process was traumatic for at least three years, and its repercussions may still be seen in the careers of individual engineers.

The energy crisis of the mid-1970s was another example of engineers responding rapidly and effectively to new conditions—from the design of fuel-efficient automobiles and energy-saving devices of all kinds to the development of alternative fuel sources and processes. In one aerospace company, engineers who had been working on the design of spacecraft life-support systems turned their abilities to the design of energy-saving systems for company buildings (Report of the Panel on Engineering Employment Characteristics).

On a profession-wide basis, there are a number of features of the engineering community that facilitate the response to changing demand, apart from the cross-disciplinary movement just described. An important resource is engineering service contractors, either individual or corporate; they tend to have a highly flexible staffing structure that lends itself to versatility and rapid changes in size. Upgrading of technicians or technologists from within a company staff represents another important adaptive response.

However, these adaptational mechanisms cannot completely solve the problem of rapidly changing demand. Their success in doing so on a broad scale tends to obscure the significant problems encountered on an individual scale—particularly when what is involved is the termination of large federal R&D programs. For one thing, severe individual hardships are brought about through career dislocation. There is also the question of whether the nation can afford the diminished utilization of technical resources that takes place when such dislocations occur.

Retraining programs offered by industry or government are of course one solution to this problem. Certain new emphases in the undergraduate engineering curriculum will help considerably (see the following section). However, the committee concludes that effective continuing education throughout a career holds the greatest promise for keeping engineers professionally flexible enough to anticipate and avoid great harm from technological obsolescence and changing demand.

Adaptability of the Engineering Organization

Adaptability of engineers is only one side of the equation governing engineering effectiveness. Although the committee did not look closely at the utilization of engineers from a managerial standpoint, many findings suggest that this is a very important issue. The ways in which engineering resources are allocated and managed within an organization appear to have an enormous bearing on the effectiveness of engineering practice in the United States. Management practices that foster an atmosphere in which creativity and innovation are encouraged can tap those potentials in their engineering staffs.

Accordingly, there is a need for corporations and government agencies to examine the relationship between their engineering management practices and general management goals. Attention to these issues would have great implications for the effectiveness of an organization.

FINDINGS, CONCLUSIONS, AND RECOMMENDATIONS

1. Between 1969 and 1982 the number of engineers in the United States nearly doubled, rising from 800, 000 to about 1.6 million. Some 75 percent of engineers work in industry and business—predominantly in the manufacturing industries (aerospace, 13.85 percent; commercial R&D, 12.1 percent; computers, 9.2 percent; and electrical machinery, 7.0 percent).
2. The federal government is highly dependent on engineering talent for many of its activities: About 6 percent of all engineers are employed directly by the government, and there is a higher proportion of engineers in the total government work force than in any other sector. Yet civil service regulations make it difficult for the federal government to compensate engineering employees at most experience levels and in most disciplines in a competitive fashion relative to industry.

   In view of the strong direct dependency on engineering talent for many of its most important activities, the federal government should review its compensation policies to ensure that it can competitively recruit and maintain a high-quality engineering work force.
3. The federal government has become a dominant user of engineering goods and services throughout the economy, employing (directly or indirectly) approximately 30 percent of the engineering work force and driving a large share of the nation's R&D.
4. Data indicate that there are far fewer technicians and technologists in the work force than there are engineers. The committee was initially concerned that this apparent weakness in engineering support implied an inefficient use of engineering resources. However, the committee found that there is a built-in asymmetry in the data for these groups. That is, many technicians and most technologists define them

selves in surveys as engineers, and many engineers do technician-level work. The occupational structure is thus not as top-heavy as it would appear to be.

Because the system appears to find the most appropriate balance through market mechanisms, there is no need at the present time to take action to alter the technician/technologist/engineer balance. However, periodic monitoring of this balance would be advisable.

5. There is a recurrent perception of discrimination against female faculty members in assignment of teaching responsibilities, in selection for research teams, and in granting tenure.

   College administrators should make a candid assessment of the attractiveness of academic life for women faculty members and, if negative aspects such as these are found, they should take firm steps to eliminate them.
6. Based on panel survey findings, industry generally believes that there is an upward trend in the quality (i.e., technical and/or intrinsic ability) of recent engineering graduates. However, most companies find that the contemporary graduate lacks the ability to step into a job and become immediately productive. Often six months to a year of additional training is required to acclimate the person to the requirements of the job. Key shortcomings here are skills in communication, group interaction (teamwork), and technical project management.
7. With the exception of short-term problems in certain industries, the committee found no evidence of an overall imbalance in supply and demand for engineers. These problems appear to be recurrent and eventually self-correcting (relying on market forces). However, the flexibility and responsiveness of the educational system is a critical factor.
8. Given present limitations in our ability to forecast economic trends and other national and international factors, it is impossible to design systems for predicting or managing supply and demand for engineers in any meaningful way.
9. The engineering educational institutions have proven to be remarkably adaptable over a long period of time, and individuals have been generally flexible in responding to change—although spot shortages and individual hardship have not been entirely avoided. Despite numerous stresses the system continues to function reasonably well today.

   No actions should be taken that would fundamentally alter the functioning of the engineering system. However, serious problems of support, of curricula, of policy and practice must be addressed if that adaptability and flexibility are to be maintained.

10. There are serious concerns about the dislocation of engineers that takes place when major changes in demand occur. Often, it is shifts in government funding for defense that drives these changes. Such events cause considerable stress for individuals and within disciplines. They also result in inefficient use of engineering resources. The committee finds that effective continuing education throughout a career holds great promise for keeping engineers flexible enough to anticipate and avoid great harm from technological obsolescence and changing demand.
11. The utilization of engineers from a managerial standpoint is an important issue. Management practices that foster an atmosphere in which creativity and innovation are encouraged can tap those potentials in their engineering employees. Thus there is a need for corporations and government agencies to examine the relationship between their engineering management practices and general management goals.

References

Bureau of Labor Statistics. 1983. National Survey of Professional, Administrative, Technical, and Clerical Pay.

College Placement Council. 1984. College Placement Council Salary Survey, No. 2 (March).

Engineering Manpower Commission. 1983. Engineering and technology enrollments—Fall 1982. Pt. I: Engineering. Washington, D.C.: AAES.

Engineering Manpower Commission. 1984a. Engineering and technology degrees, 1983. Pt. III: By curriculum. Washington, D.C.: AAES.

Engineering Manpower Commission. 1984b. Engineering Manpower Bulletin, No. 73 .

Geils, J. 1982. The faculty shortage: The 1982 survey. Engineering Education (November), pp. 147–158.

IEEE United States Activities Board. 1984. Profile of U.S. IEEE Women Members: Their Salaries, Demographics, Attitudes Toward the Workplace and Professional Status. New York: IEEE.

National Science Board. 1983. Science Indicators: 1982. Washington, D.C.: National Science Foundation.

National Science Foundation. 1982a. Changing Employment Patterns of Scientists, Engineers, and Technicians in Manufacturing Industries: 1977–80. Washington, D.C.: National Science Foundation.

National Science Foundation. 1982b. University-Industry Research Relationships: Myths, Realities, and Potentials. Washington, D.C.: National Science Foundation.

National Science Foundation. 1983. U.S. Scientists and Engineers. Washington, D.C.: National Science Foundation.
National Science Foundation. 1984. Women and Minorities in Science and Engineering. Washington, D.C.: National Science Foundation, Division of Science Resource Studies.
Office of Technology Assessment. 1984. Computerized Manufacturing Automation: Employment, Education, and the Workplace. Washington, D.C.: U.S. Government Printing Office.
Report of the Panel on Engineering Employment Characteristics, in preparation.
Report of the Panel on Engineering Interactions With Society, in preparation.
Report of the Panel on Graduate Education and Research, in preparation.

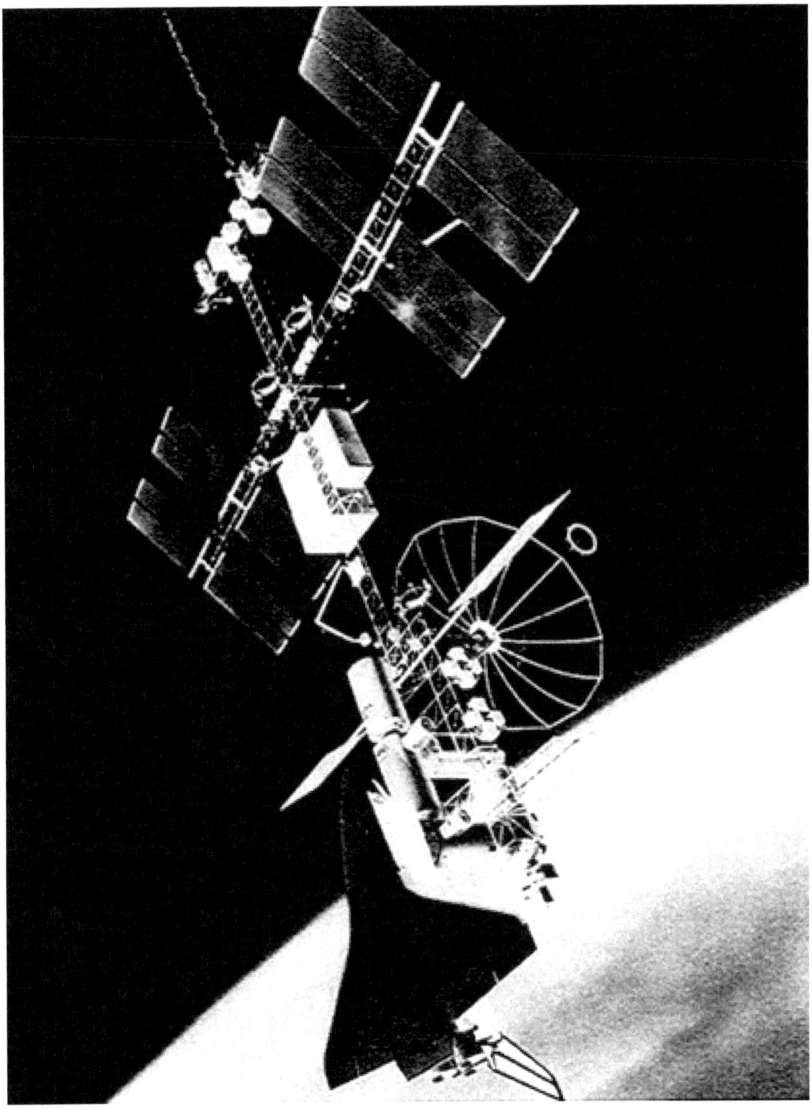

A LOOK AT THE FUTURE

Photograph on overleaf courtesy of the National Aeronautics and Space Administration

6

Engineering's Future: Requirements for a Changing Environment

THE YEAR 2000: WHAT WILL THE ENGINEERING ENVIRONMENT BE LIKE?

Looking forward to the year 2000 (or to any future year), the committee believes that it is the goal of those who are responsible for the education of engineers and the organization of engineering's effort to ensure that economically or socially beneficial products or services are not delayed or denied to society because of an inadequate engineering establishment in the United States.

Likely Characteristics

One way—perhaps the best way—to gauge the means and mechanisms by which engineers are educated and utilized is to begin by identifying likely general differences between the United States of today and the United States in the year 2000. We may then consider how the existing means and mechanisms can be adjusted to ensure that the engineering community will provide effective, efficient support for such likely and evolutionary changes.

Assuming that there is no global conflagration during the next 15 years, the committee believes that the United States of 2000 will very likely be characterized in the following ways:

- The time horizons over which U.S. industry seeks to maximize its profits will likely be longer than those of today.

- While suppliers of capital will often take a longer view of the performance of businessmen in allocating financial resources, there will nevertheless be significant shortages of capital for at least some industries and firms.
- The United States will increasingly be an integral part of a truly global economy, with international trade as a growing component of United States economic activity. Generally increased interindustry and intraindustry competition will characterize this global economy.
- Because of developments in defense, energy, space, and other areas, government demand for engineering goods and services [both direct and indirect] will increase significantly in proportion to other sources of demand.
- Whether or not energy materials remain relatively scarce, the economy of 2000 and beyond will face raw materials shortages (in some cases chronic shortages).
- Scientific discoveries and technology development will continue to occur at a rapid rate. This process will make possible the seminal, revolutionary advances that create new industries; it will also give engineers a larger menu of technical tools and options for existing tasks.
- At the same time the number of engineering tasks that do not require cutting-edge engineering will continue to increase—as evidenced by the growing need to maintain, rehabilitate, and operate the nation's aging infrastructure.

Before elaborating upon the impacts and implications of the foregoing likely differences between 1985 and 2000, it is important to reiterate that they have been postulated here to permit us to arrive at judgments concerning the education and utilization of engineers in the United States. Other changes may prove to be equally important; nor will all of those described necessarily be seen. However, if the available means and mechanisms for educating engineers and allocating their services can cope satisfactorily with the changes outlined, they should be capable of dealing with virtually anything the future has in store for the United States.

This assertion is predicated on the assumption that many of the recommendations in this report have been heeded and implemented. That is, because of the focus on engineering science and fundamentals, the educational system will have produced thoughtful, flexible engineering talent. The managers of both government and private organizations will fully understand that engineering effectiveness depends to a great extent on how well the engineering effort is managed. As a result, they will have devised means and mechanisms for organizing engineering resources in such a way as to meet, with acceptable efficiency, the demands placed upon the engineering community.

Impacts and Implications

In the United States of the mid-1980s, the time horizons of managers in most American companies and industries are short because of the pressure for quarter-to-quarter earnings improvements emanating from institutional investors and corporate stockholders. Because this situation does grievous long-term harm to the United States—especially in markets where much of the competition is from abroad—it is reasonable to expect that a more rational approach will emerge, either through government action or through changed attitudes on the part of investors or both. Hence, the time horizons of managers can be expected to lengthen substantially.

The implications for engineering are significant. For example, there will be increasing emphasis on capital-intensive solutions to production problems. Product quality can be improved as investments in plant and equipment as well as in the education and training of employees become not only tolerable but required. In turn, the demand for technology-intensive capital goods will be greater, and the range of engineering disciplines required to meet that demand will certainly be very broad. However, the demand will not be uniform across the spectrum of engineering at any point in time. Consequently, engineers capable of working in adjacent disciplines will function better than those who are more narrowly educated.

Public/private sector versatility. Similarly, the further growth of government demand for engineering goods and services will create a need and an advantage for engineers who are capable of functioning in both the public and private sectors. A basic requirement here is that such engineers must understand the different management objectives of these two sectors.

Private-sector objectives are driven by competitive markets, while public-sector objectives are driven by political and public concerns. Thus engineers in each sector place different degrees of emphasis on the common engineering concerns for innovation, cost containment, productivity, safety, consumer satisfaction, and protection of the environment (Report of the Panel on Engineering Interactions With Society).

The committee concludes that sensitizing students to these basic differences in the servicing of the public and private sector is of considerable importance. If engineers know and can recognize the differences, regardless of the occupational environment in which they find themselves, they will be able to understand how and why approaches differ. In this way they will be much better able to move between public-sector and private-sector career opportunities. In addition, better understanding by engineers in each sector of the basic objectives of the other sector will yield better, more economical products and services in both sectors.

Shortages of resources. Throughout most of the history of engineering in the United States, engineers have been educated and oriented to deal with situations characterized by a sufficiency, if not a surplus, of resources for use in the task at hand. Since the energy crisis of 1973, there have been modest attempts to introduce into engineering curricula materials that suggest engineers may have to face shortages of one resource or another. Generally, the emphasis in this regard has been on energy. Since it is likely that both spot and chronic shortages of materials of various kinds will increasingly characterize the economy of the future, it is increasingly important for both engineering education and practice to reflect that fact. Students will need to learn how to deal with shortages in resources so that they may take explicit account of them when performing engineering functions in the economy.

Of all the resources that will periodically be scarce in the future, none can be so predictably forecast as shortages related to capital—to financial resources. Expensive capital (i.e., capital in short supply) will severely affect building and construction, venture capital availability (and thus the number of start-ups), and modernization and expansion efforts. Since capital constraints will be a very real aspect of the operational environment of the future, it will be essential for students to understand the impact of these constraints on planning and design.

Global economy. It is not enough for engineers to be trained and employed in such a way that only U.S. markets and conditions are taken into account. Inevitably, the United States must become increasingly bound up in the world economy. This means that the practice of engineering will have to take account of what foreign markets require and will accept. (An obvious example is the growing importance of standards and interchangeability on a worldwide basis.) It also means that international competition between the engineering work forces of different countries will intensify. This is not a subject that the commit tee was able to examine in detail, although it clearly has major implications for the future.[1]

One important implication of the global economy is that it requires sensitivity to regional and cultural differences and their impact on worldwide demand for engineering goods and services. Engineers will also need to appreciate the financial, political, and security forces at play internationally. The nontechnical components of engineering education ought to include exposure to these aspects of contemporary engineering.

In this context, communication among U.S. engineers and engineering-based companies is crucial if the United States is to maximize the net benefits it derives from participation in international trade and in other aspects of the global economy. The engineering community ought to be prepared to promote open communication of this kind, especially with regard to the goods and services that the world (and not merely the United States) requires and is prepared to accept.

Rapid scientific and technological change. Science and technology have been invaluable contributors to the expansion and success of the U.S. economy. This will be no less true in the foreseeable future than in the past. Here again, the implications for engineering education and utilization are very great. Indeed, engineering practice has already been undergoing a revolution over the past several years. New engineering tools based on the computer, such as computer-aided design and computer-based workstations, are part of this revolutionary change. New methods such as simulation and modeling are driving engineering activity in the direction of greater abstraction—more mathematical analysis, less experimentation.

There is no apparent slowdown in this revolution in practice. In fact, it will continue to accelerate, and will gain further impetus from additional progress in such technologies as composite materials, expert systems, and supercomputers. With their creative and productive capabilities greatly enhanced through the use of such tools and methods, engineers in every discipline will be able to turn increasingly from the

[1] Differences in the roles and responsibilities of engineers in different countries, as well as a lack of adequate data, make direct comparisons difficult. Some sources of reference in this area are Mintzes, 1982; Mintzes and Tash, 1984; National Research Council, 1984; National Science Board, 1983; Office of Technology Assessment, 1983; Office of Technology Assessment, 1984; and Secretary of State for Industry, 1980.

mechanical to the conceptual. Many of them will be less involved in the performance of conventional or routine engineering work and more involved in the formulation of ideas, in making choices.

Consequently, an increasingly important element of engineering education will be to teach engineers to approach problems—that is, how to ask the right questions and know the dimensions of responsive answers—even when the details of a project are entirely new with regard to materials or processes, environmental issues, or markets.

The options and opportunities based in changing scientific and technological possibilities are vast. Therefore, engineers need to be well rounded in science and increasingly knowledgeable about scientific advances that have promise for supporting engineers facing specific, related project responsibilities and objectives. Engineers should also be equipped to play a substantial role in the various processes of technological innovation that are essential to the well being of the United States, both in civil and military contexts. Engineers who understand and appreciate the scientific and technological underpinnings of the products and processes with which they are involved can participate to the utmost in innovation processes, especially if they have also been educated in the fundamentals of innovation.

Because of its focus on research, engineering doctoral study is at present one of the best ways to acquire a strong orientation toward scientific and technological innovation. The Ph.D. will continue to be valuable, both to the profession and to the individual degree holder. In the short term, there will be a great need for more of the best engineering students to obtain the doctoral degree and become engineering professors. Given the expected increase in emphasis on research and innovation in most industries, in the long term it will be beneficial for the nation as a whole if more United States residents of the highest academic caliber choose to continue on for the Ph.D.

Notwithstanding the expansion of scientific discoveries and technological possibilities, society will continue to require substantial—even growing—engineering services of a less advanced nature. This is especially apparent with regard to the expanding need to maintain the aging plant and equipment found in both the public and private sectors.

Need for economic awareness. Despite the anticipated involvement of government in the U.S. economy, in the private sector domestically, and in international trade generally, heightened competition on both the interindustry and intraindustry levels can be safely projected. This implies that engineers must establish and maintain great sensitiv ity to the economic aspects of engineering; these cannot be treated as subordinate issues. To do so would jeopardize the usefulness and value of individual engineers; it would also produce engineering results that do not serve the interests of the U.S. economy to the extent that they can and should.

MEANS AND MECHANISMS FOR ADAPTING SUCCESSFULLY

What is needed to enable the engineering community to adapt to these likely future conditions (still assuming that the ability to cope with those conditions implies an ability to cope with any likely future)? The foregoing section as well as earlier sections have identified a number of different characteristics and strengths the engineering community must acquire to ensure that the United States maintains its relative position in the world and that engineering continues to meet the nation's needs. Many of these requirements relate to the kind of education that engineers receive before entering practice. Others relate to their subsequent responsibilities as professional men and women. They are drawn together here from various sections of the report in order to bring into clearer focus the range of requirements that the engineering community will need to address effectively if it is to meet the demands that the future will place upon it.

Curriculum Requirements

Broad engineering education. Of foremost importance is the ability to impart a strong, diversified engineering education—one incorporating depth of specialization as well as breadth, with a strong grounding in the fundamentals. To the extent that there has been movement toward the concept of basic engineering and general education, followed by specific study in the engineering field, the committee encourages that trend. Dual-degree and other alternative curricula should be examined to see whether they can expand the benefits of this approach.

Stronger nontechnical education. Related to the broader engineering education urged by the committee is the need for better general education of engineers. Exposure to course work in the humanities, arts, and social sciences over an extended period of time (i.e., beyond just the freshman and sophomore years) offers many advantages in molding the contemporary engineer. Among the most tangible of these is an improved facility for communication, both written and oral. Sev eral recent authoritative reports have stressed the importance of the humanities, in particular, in shaping a young man's or woman's judgment and system of values (see, for example, Bennett, 1984).

Greater exposure to the world of ideas in general renders an engineer better equipped to function on an equal footing, both professionally and socially, with corporate peers and managers of varied educational backgrounds. In the real world of the workplace, such fluency is important in enabling engineers to represent effectively the interests, needs, and objectives of the engineering department within the organization. Finally, education of this type prepares an engineer to better anticipate, understand, and adapt to the new and changing conditions—whether they be social, economic, cultural, or political—that will affect technology development in the global marketplace of the future.

Exposure to computer technology. It is certain that the computer will become pervasive in the practice of engineering, both as a tool for performing the engineering job itself (e.g., in design) and as a medium for carrying out many other necessary activities (e.g., communication, recordkeeping, and reporting). Consequently, engineering education in every discipline must include some exposure to computer science and programming. Computers are at present a more central feature of the educational experience in some disciplines than in others—in, say, electrical engineering than in civil engineering. Budgetary constraints are certainly a factor here in most schools. However, a goal of engineering school administrators should be to see that every department has access to the available computer resources.

Orientation to the realities of the work world. The context in which engineering work is carried out is changing in a number of ways, as described in the previous section. Many of these changing features of the environment have implications for engineering curricula, apart from those already discussed. Increasingly frequent and severe shortages of materials of various kinds, for example, will require that engineering students learn how to deal with resource shortages as one type of constraint on design. Another type of resource constraint is shortages of capital, which will likely be a frequent consideration for the foreseeable future. Students must likewise be able to understand and deal with the impact of this constraint on planning and design.

A third requirement derives from the expected further growth in government use of engineering resources. The engineering educational process should make students aware of the differing objectives and driving forces that, in general, characterize engineering in the public and private sectors. Such an awareness is important for engineers in either sector as the interaction between sectors increases and especially as the flow of engineers between sectors increases. Finally, an awareness of the different cultural objectives and forces characterizing different regions of the world market (at least in general) will help engineers to have a better sense of the dynamics and requirements of international competition for these markets. Such knowledge, specific to the realities of the world market for engineering products, could form one of many links between the technical and nontechnical components of engineering education.

Personal career management. Many of the points made in the section on characteristics of the future highlighted the need for career adaptability. It is generally not part of engineering curricula—undergraduate or graduate, formal or informal—to provide engineers with the insights necessary to promote their ability to manage their own careers in any long-term sense. This is a great shortcoming in engineering education. Certainly if engineers are made aware of the options and opportunities they will face in the future—as well as the problems and pitfalls—the allocation of resources both to and within engineering will be far more efficiently carried out than would otherwise be the case. The ability to actively and intelligently manage one's engineering career would benefit not only in individuals but, in the aggregate, the nation as a whole.

The foregoing represents a considerably long list of topics and new educational emphases recommended for inclusion within the undergraduate engineering curriculum. Yet, as was discussed in chapter 4, it is difficult to provide even the cursory exposure to nontechnical subjects currently required by most schools, within a four-year program. There is frequent pressure to reduce even that small requirement in order to satisfy the demand for greater technical content. The committee is well aware that to expect the current curriculum to be expanded to accommodate greater breadth and depth of engineering study as well as more nontechnical educational and orientational subjects would be naive. Yet, these educational components will be increasingly necessary if American engineers and engineering are to maintain the flexibility and resiliency that the future environment will demand.

On that basis the committee concludes that some restructuring of the undergraduate curriculum will have to occur. What form it will take will vary from school to school. Some of the material can be woven into existing courses by changing the way in which courses are taught. Greater flexibility in course requirements is another conservative approach, allowing more courses of this type to be taken as electives. Five-year programs, including dual-degree programs, address the problem of too-limited time more directly. The concept of professional engineering study following general education, as in the medical profession, has even been proposed. In any case, engineering schools will have to examine their own circumstances very closely, with a view to determining how these important educational needs can begin to be addressed.

Requirements for the Professional Career

Greater management skills. Regardless of an engineer's field of work, an important characteristic will be the possession of greater management skills—in the sense of technical project management and management of the engineering task at hand—than have been seen among engineers in the past. The ability to work in teams and to relate to other functions of the larger organization (e.g., marketing and finance) is an essential element of these skills. In a more competitive world, it will be advantageous if technical activities are managed competently and directly by technically oriented people.

Broader education in both technical and nontechnical fields, as called for in the previous section, will be important in preparing an engineer intellectually for the complex demands of project management. Nevertheless, the essential temperamental and experimental preparation for those responsibilities is gained not in the classroom but in the workplace. In the absence of specific on-the-job training for this purpose, personal initiative on the part of the individual engineer will continue to be necessary for gaining competence in these highly interpersonal and sometimes political skills. Early work experience, whether acquired through cooperative education, summer employment, or some other route, can also be a primary source of these practical skills. In addition, work experience exposes the budding engineer to documentation, reporting procedures, and other practical aspects of basic engineering project and task management.

One intangible but important need in these challenging times is for the development of a stronger sense among engineers of their professional role and its responsibilities. Professional ethics is a part of this responsibility and is part of the impetus toward a broader education of engineers (Christiansen, 1984; Report of the Panel on Support Organizations and the Engineering Community). Public criticism of engineering and technology has abated in recent years, but from it the engineering community has learned an important lesson. That is, the innovation and management of complex technical systems involves a consideration of social preferences and impacts as well as technical knowledge and skill. Translating such considerations into corporate policy has been perennially difficult. But if more engineers become sensitized to the social ramifications of their work, their viewpoints will represent a formidable force within industry (Report of the Panel on Engineering Interactions With Society).

Career effectiveness. The effectiveness of engineers depends upon their knowledge and capabilities. Those characteristics, in turn, are a function of experience, training, and—almost as importantly—the management approaches that prevail within the organization. The organizational philosophy toward continuing education, in particular, can greatly facilitate the effectiveness of engineering employees throughout their careers. It is estimated that because of the rapidity of technological change, an engineer who does not learn while working now has a useful life in practice of only about 10 years. Easier access to technical education and training throughout their careers will be necessary if engineers are to keep current in their field and keep abreast of developments in other fields. Such continuing education should include timely access to effective retraining programs.

However, formal continuing education alone is not enough. Only about a roughly estimated 5 percent of an engineer's continuing educational opportunities are of this type. The other 95 percent consist of a wide range of informal experiences including on-the-job learning, conferences, seminars, short courses, and so forth (Report of the Panel on Continuing Education). Nor is it enough for management to be willing to make these opportunities available. Engineers as individuals must have the personal motivation necessary to take advantage of these learning opportunities. A healthy respect for the career effects of obsolescence is certainly one basis for that motivation. But it must also be based on a clear understanding of broad world and national economic and technological trends and on a confidence in one's ability to maintain individual competency and marketability through individual initiative.

FINDINGS, CONCLUSIONS, AND RECOMMENDATIONS

1. Likely characteristics of the engineering environment in the year 2000 include longer time horizons for profit-taking in industry, shortages of capital and resources (both energy and materials), a global economy, with increased intra-and interindustry competition, increased government demand for engineering goods and services, continued

high rate of scientific discovery and technology development, and an increased requirement for nonadvanced engineering tasks.

2. Given anticipated growth in government demand for engineering resources, sensitizing students to basic differences in the servicing of the public and private sectors is of considerable importance.
3. Because it is likely that both spot and chronic shortages of materials (as well as energy and capital) will characterize the economy of the future, it is important for engineering education and practice alike to reflect those constraints.
4. In the context of an increasingly global economy, sensitivity to cultural and regional differences will be important qualities for engineers to acquire. Engineers will also need to appreciate the financial, political, and security forces at play internationally. Communication among U.S. engineers and engineering-based companies regarding the nature of international demand for goods and services will be crucial.

   The nontechnical components of engineering education ought to include exposure to these aspects of contemporary engineering. In addition, the engineering community should strive to ensure open communication on these matters among engineers and companies the world over.
5. Continuing scientific discovery and technology development will give further impetus to a revolution in engineering practice. With the use of new tools and methods the work of many engineers will become increasingly abstract, involving formulation of ideas and choosing among development options. Therefore engineers will need to be able to deal with problems in unfamiliar contexts, they will need to be knowledgeable about scientific advances generally, and they should understand the fundamentals of innovation.
6. With heightened competition among and between industries, both domestically and internationally, engineers must establish and maintain great sensitivity to the economic aspects of engineering.
7. If United States engineers are to be adequately prepared to meet future needs, then the undergraduate engineering curriculum must emphasize broad engineering education, with strong grounding in fundamentals and science. In addition, the curriculum must be expanded to include greater exposure to a variety of nontechnical subjects as well as work-orientational skills and knowledge. To accomplish this expansion will require restructuring of the standard four-year curriculum by various means.

   Engineering schools will have to examine their existing curriculum and their particular circumstances closely in order to ascertain how the curriculum can best be restructured to address these important educational needs.

In addition, the committee has recommended that the National Science Foundation fund a pilot group of engineering schools to evaluate dual-degree and other alternative educational programs experimentally (see chapter 4, recommendation 15). The results of this experimental program are likely to be quite relevant to the question of curriculum structure and nontraditional content. Participating schools (and engineering school administrators generally) should examine the results from this standpoint.

8. Successful adaptation to future conditions will require that practicing engineers develop a number of attributes. These include greater technical project management skills, a stronger sense of professional role and responsibilities, and a strong orientation toward maintenance of effectiveness through continuing education.

References

Bennett, W. J. 1984. To Reclaim A Legacy: A Report on the Humanities in Higher Education. Washington, D.C.: National Endowment for the Humanities.

Christiansen, D. 1984. The issues we avoid. IEEE Spectrum, 21(6), p.25.

Mintzes, J. 1982. Scientific and Technical Personnel Trends and Competitiveness of U.S. Technologically Intensive and Critical Industries. Prepared for Division of Policy Research and Analysis, National Science Foundation.

Mintzes, J. and Tash, W. 1984. Comparison of Scientific and Technical Personnel Trends in the U.S., France, West Germany, and the United Kingdom Since 1970. Prepared for Division of Science Resource Studies, National Science Foundation.

National Research Council. 1984. Labor Market Conditions for Engineers: Is There a Shortage? Proceedings of a Symposium held by the Office of Scientific and Engineering Personnel. Washington, D.C.: National Academy Press.

National Science Board. 1983. Science Indicators: 1982. Washington, D.C.: National Science Foundation, pp. 2–37.

Office of Technology Assessment. 1983. International Competitiveness in Electronics. Washington, D.C.: U.S. Government Printing Office.

Office of Technology Assessment. 1984. Computerized Manufacturing Automation: Employment, Education, and the Workplace. Washington, D.C.: U.S. Government Printing Office.

Report of the Panel on Continuing Education, in preparation.

Report of the Panel on Engineering Interactions With Society, in preparation.

Report of the Panel on Support Organizations and the Engineering Community, in preparation.

Secretary of State for Industry. 1980. Engineering Our Future: Report of the Committee of Inquiry Into the Engineering Profession. London, England: Her Majesty's Stationery Office.

APPENDIX A

SUBCOMMITTEE AND PANEL MEMBERS, CONSULTANTS, AND LIAISON REPRESENTATIVES

SUBCOMMITTEE ON EDUCATIONAL SYSTEMS

JORDAN J. BARUCH, Chairman; President, Jordan J. Baruch Associates, Washington, D.C.

EDMUND T. CRANCH, President, Worcester Polytechnic Institute

JOHN D. KEMPER, Professor, Department of Mechanical Engineering, University of California at Davis

EDWARD T. KIRKPATRICK, President, Wentworth Institute of Technology

ERNEST S. KUH, Professor, Electrical Engineering and Computer Science, University of California at Berkeley

ROBERT M. SAUNDERS, Professor, School of Engineering, University of California at Irvine (Chairman, Board of Governors, AAES, 1983)

CHARLES E. SCHAFFNER, Executive Vice-President, Syska & Hennessy, New York, N.Y.

MORRIS A. STEINBERG, Vice-President, Science, Lockheed Corporation

SHEILA E. WIDNALL, Professor, Aeronautics and Astronautics, Massachusetts Institute of Technology

PANEL ON UNDERGRADUATE EDUCATION

EDMUND T. CRANCH, Chairman; President, Worcester Polytechnic Institute

EUGENE M. DELOATCH, Dean, School of Engineering, Morgan State University, Baltimore, Maryland

DONALD G. GLOWER, Dean, College of Engineering, Ohio State University

WILLIAM R. GROGAN, Dean, Undergraduate Studies, Worcester Polytechnic Institute

CHARLES E. SCHAFFNER, Executive Vice-President, Syska & Hennessy, New York, N.Y.

WILLIAM R. UPTHEGROVE, Regents Professor of Engineering, School of Aerospace, Mechanical and Nuclear Engineering, University of Oklahoma

SHEILA E. WIDNALL, Professor of Aeronautics and Astronautics, Massachusetts Institute of Technology

DONALD TRITSCHLER, consultant, Shrewsbury, Massachusetts

PANEL ON TECHNOLOGY EDUCATION

EDWARD T. KIRKPATRICK, Chairman; President, Wentworth Institute of Technology

JOHN D. ANTRIM, General Manager, Certification Programs, National Society of Professional Engineers

STEPHEN R. CHESHIER, President, Southern Technical Institute

RICHARD A. KENYON, Dean, College of Engineering, Rochester Institute of Technology

LAWRENCE J. WOLF, Dean, College of Technology, University of Houston

DONALD TRITSCHLER, consultant, Shrewsbury, Massachusetts

PANEL ON CONTINUING EDUCATION

MORRIS A. STEINBERG, Chairman; Vice-President, Science, Lockheed Corporation

RALPH T. DOSHER, JR., Manager, Corporate Education, Texas Instruments, Inc.

ROD HANKS, Director, College Relations and Technical Development, Lockheed Corporation

ROBERT A. HOPSTADER, Manager, Education and Development Unit, Exxon Research and Engineering Company

HAROLD G. KAUFMAN, Associate Professor of Management, Director of Research Programs in Science, Technology and Human Resources, Polytechnic Institute of New York

RUSSELL R. O'NEILL, Professor, School of Engineering and Applied Science, University of California at Los Angeles

BERNARD M. SALLOT, President, Advanced Technologies Group Services, Farmington Hills, Michigan

ROBERT F. BORUCH, consultant, Psychology Department, Northwestern University

RICHARD DRECHSLER, consultant, New York, N.Y.

KENNETH M. REESE, consultant, Washington, D.C.

PANEL ON GRADUATE EDUCATION AND RESEARCH

JOHN D. KEMPER, Chairman; Professor, Department of Mechanical Engineering, University of California at Davis

JAMES F. MATHIS, Vice-President, Science and Technology, Exxon Corporation

IRENE C. PEDEN, Professor, Department of Electrical Engineering, University of Washington at Seattle

JOSEPH E. ROWE, Executive Vice-President, Research and Defense Systems, Gould Inc.

LELAND J. WALKER, Chairman of the Board, Northern Engineering & Testing

PANEL ON INFRASTRUCTURE DIAGRAMMING AND MODELING

ERICH BLOCH, Chairman; Vice-President, IBM Corporation (now Director, National Science Foundation)

KAREN C. COHEN, Principal Research Associate, Center for Advanced Engineering Study, Massachusetts Institute of Technology

ARNOLD R. ESHOO, Program Manager, Technical Resource Statistics, Technical Personnel Development, IBM Corporation

JOHN W. GEILS, Director, AAES/ASEE Faculty Shortage Project (AT&T, Ret.)

DEAN GILLETTE, Executive Director, Corporate Studies, AT&T, Bell Labs (now Professor, Engineering Department, Harvey Mudd College)

L. OWEN HILL, Manager, Site Resource Management, IBM East Fishkill, Hopewell Junction, N.Y.

WILLIAM K. LEBOLD, Director, Educational Research and Information Systems, Purdue University

HAROLD T. SHAPIRO, President, University of Michigan

PATRICK J. SHERIDAN, Executive Director, EMC, American Association of Engineering Societies

DONALD G. WEINERT, Executive Director, National Society of Professional Engineers

ROBERT P. STAMBAUGH, consultant, Quechee, Vermont

MICHAEL CROWLEY (liaison representative), Staff Associate, Division of Science Resource Studies, National Science Foundation

DANIEL E. HECKER (liaison representative), Office of Economic Growth and Employment, United States Department of Labor, Bureau of Labor Statistics

PANEL ON ENGINEERING EMPLOYMENT CHARACTERISTICS

FRED W. GARRY, Chairman; Vice-President, Corporate Engineering and Manufacturing, General Electric Company

JOHN D. CULLEN, Senior Vice-President, Engineering, E. I. DuPont de Nemours & Company

DANIEL C. DRUCKER, Dean of Engineering, University of Illinois at Urbana (now Graduate Research Professor of Engineering Sciences, University of Florida at Gainesville)

HELEN GOULDNER, Dean, College of Arts and Science and Professor of Sociology, University of Delaware

M. EUGENE MERCHANT, Principal Scientist, Manufacturing Research, Cincinnati Milacron, Inc. (now Director, Advanced Manufacturing Research, Metcut Research Associates Inc.)

LAWRENCE M. MEAD, JR., Senior Management Consultant, Grumman Aerospace Corp.

GORDON H. MILLAR, Vice-President, Engineering, Deere and Company (now retired)

RICHARD J. REDPATH, Vice-President, Ralston Purina

JUDITH A. SCHWAN, Assistant Director, Research Laboratories, Eastman Kodak Company

STEPHEN TUCKER, Program Manager, Edison Engineering, General Electric Company

BERNARD CULLEN, consultant, McBer and Company

KENNETH M. REESE, consultant, Washington, D.C.

JEANNE CARNEY (liaison representative), Staff Specialist, OUSDRE (R&AT), Office of Research and Lab Management, the Pentagon

PANEL ON REVIEW OF CURRENT CONCERNS AND RESPONSES

JOHN W. GEILS, Chairman; Director, AAES/ASEE Faculty Shortage Project (AT&T, Ret.)

GEORGE E. DIETER, JR., Dean, College of Engineering, University of Maryland

DAVID F. THOMAS, consultant, ManTech International Corp.

PANEL ON ENGINEERING INTERACTIONS WITH SOCIETY

GEORGE S. ANSELL, Chairman; Dean of Engineering, Rensselaer Polytechnic Institute (now President, Colorado School of Mines)

THOMAS P. CARROLL, Professor, Science & Technology Studies Division, Rensselaer Polytechnic Institute

SAMUEL FLORMAN, Vice-President, Kreisler, Borg, Florman Construction Company

AARON J. GELLMAN, President, Gellman Research Associates, Inc.

MELVIN KRANZBERG, Callaway Professor of the History of Technology, School of Social Studies, Georgia Institute of Technology

LAWRENCE M. MEAD, JR., Senior Management Consultant, Grumman Aerospace Corp.

M. EUGENE MERCHANT, Principal Scientist, Manufacturing Research, Cincinnati Milacron, Inc. (now Director, Advanced Manufacturing Research, Metcut Research Associates, Inc.)

ARTHUR L. DONOVAN, consultant, Professor and Director, Center for the Study of Science and Society, Virginia Polytechnic Institute and State University

COURTLAND S. LEWIS, consultant, Washington, D.C.

PANEL ON SUPPORT ORGANIZATIONS AND THE ENGINEERING COMMUNITY

FRANCIS E. REESE, Chairman; Senior Vice-President, Monsanto (now retired)

DENNIS CHAMOT, Associate Director, Department for Professional Employees, AFL/CIO

LLOYD A. DUSCHA, Deputy Director of Engineering and Construction, U. S Army Corps of Engineers

FRED JEROME, Director, Scientists' Institute for Public Information, Media Resource Service

JOHN E. KING, Director, Engineering Business Operations and Administration, Douglas Aircraft Company

CHARLES A. SORBER, Associate Dean, College of Engineering, The University of Texas at Austin

DONALD G. WEINERT, Executive Director, National Society of Professional Engineers

PAULA B. WELLS, Executive Vice-President, Wells Engineers, Inc.

APPENDIX B

CONTRIBUTIONS TO THE COMMITTEE REPORT GENERATED BY PARTICIPANTS IN THE STUDY

Report of the Subcommittee on Educational Systems

Report of the Panel on Undergraduate Education

Report of the Panel on Graduate Education and Research

Report of the Panel on Technology Education

Report of the Panel on Continuing Education

Report of the Panel on Infrastructure Diagramming and Modeling

Report of the Panel on Employment Characteristics

Supplements on women in engineering and on the social content of minorities in engineering contributed by Helen Gouldner

Supplement on the role of the federal government in the education and utilization of engineers contributed by W. Edward Lear and Donald G. Weinert

Report of the Panel on Engineering Interactions with Society

Supplement on "Engineering in an Increasingly Complex Society: Historical Perspectives on Education, Practice, and Adaptation in American Engineering." Report of a Conference, July 19–21, 1983, sponsored by the committee, prepared by Arthur L. Donovan

Report of the Panel on Support Organizations and the Engineering

Community

Supplement of task force on society-at-large contributed by Fred Jerome

Supplement on the professional/technical engineering societies contributed by Donald G. Weinert

Report of the Panel on Concerns and Responses Regarding Engineering Education

INDEX